A Better Integrated Management of Disaster Risks

Toward Resilient Society to Emerging Disaster Risks in Mega-Cities

Cover illustration: Urban flood disasters on September 2000 in the Tokai Regions.

A Better Integrated Management of Disaster Risks

Toward Resilient Society to Emerging Disaster Risks in Mega-Cities

Edited by

S. Ikeda, T. Fukuzono, and T. Sato

National Research Institute for Earth Science and Disaster Prevention

TERRAPUB, Tokyo

A Better Integrated Management of Disaster Risks:
Toward Resilient Society to Emerging Disaster Risks in Mega-Cities
Edited by S. Ikeda, T. Fukuzono, and T. Sato
ISBN 978-4-88704-140-0

Published by TERRAPUB, 2003 Sansei Jiyuugaoka Haimu, 27-19 Okusawa 5-chome, Setagaya-ku, Tokyo 158-0083, Japan.
Tel: +81-3-3718-7500 Fax: +81-3-3718-4406
URL http://www.terrapub.co.jp/

Printed in Japan

Contents

Part III: Pilot studies of implementing social platform of risk management in local community: Participatory flood risk communication support system (Pafrics)

Preface

The characteristics of natural disasters in modern mega-cities have been complicated and diversified since the end of 20th century and originated not only from geo-physical and global changes of natural hazards, such as earthquakes, tsunami, floods, storms, landslides, and drought, but also from the recent structural changes of our post-industrial societies. In fact, the Japanese urban area has become more and more vulnerable to newly emerging risks of low probability but high consequence type (LPHC type), which have been induced technologically and socially in modern society. For example, the estimated damage exposed by the floods has increased five times during the past 20 years in terms of unit-exposed space (inundated area in the case of flooding). This is due to the high concentration of both population and social infrastructures in residential districts, in spite of the success in decreasing the total number of human casualties of flood disasters, thanks to the tremendous amount of investment in constructing such physical structural facilities as dams, dikes, banks, etc.

In return, we unfortunately face the soaring marginal cost of reducing disaster risk up to an acceptable level. In addition, we have such a new type of multi-disasters in modern mega-cities that a single small-scale hazard episode in an urban area might trigger a series of "catastrophic disasters" in a cascading manner under current interwoven and complicated urban water systems, such as from a small-scale river channel up to a large-scale river embankment. There seems to be many cases in which "natural disasters with totally different characteristics may happen before people forget about the previous disasters", in contrast with the famous expression "natural disasters occur when people are not thinking about them".

Until very recently, people used to take precautionary steps against storms and/or inundation disasters at their homes in accordance with the amount of rainfall or the rise of water in adjacent rivers, and would listen to radio or TV news reports that predicted the routes of typhoons, and would standby at their homes or work places. Nowadays, such self-help practices to reduce possible damage of disasters have waned, reflecting a public relience on the remarkable improvements in basic social infrastructure to prevent disasters.

Our country seems to have been transformed into a "hands-off society of leaving the management responsibility to regulatory authorities" in which people wait until they receive warning information or evacuation orders from the local government who are specialized in disaster prevention. However, when we face unexpected or surprise conditions, ordinary citizens become unable to understand the nature of risks, partly due to the complex processes of the regulatory management systems specialized in current disaster prevention schemes. They stop making their own choices regarding proactive responses to the emerging risks that have been so-called socio-technologically or socio-culturally constructed in our post-industrial society.

In fact, newcomers who have migrated and settled in newly developed lowland areas that used to be flood plains tend to be unable to inherit past disaster experiences. They tend to claim more and more "safety measures of intensified structural facilities" as well as an expansion of the "disaster relief measures" in cases where local residents become victims. But they prefer to leave management judgment regarding distinction between "safety and danger" to governmental institutions, as indicated in our recent case studies on a series of catastrophic flood disasters, such as the 2000 Tokai Floods in Nagoya and the 2004 Niigata-Fukui Floods. This also leads to another management issue of the increase in social costs to deal with the unexpected nature of risks, even by taking a proper combination of both structural and non-structural measures depending on which level of risks the community and residents may accept in long-term. Moreover, we are concerned about this kind of emerging disaster risks growing larger and continuing to increase, as suggested by the recent IPCC reports of global and regional climate changes.

Here appears an integrated risk management approach to deal with natural disasters as social risk phenomena in terms of enhancing a variety of proactive or participatory ways that governmental institutions, communities, and residents could jointly carry out proper risk management against newly emerging disaster risks. To make our modern societies sustainable in terms of our lives, socio-cultural assets, and environmental resources in this risk society, it is vital to create a kind of societal governance system resilient to unexpected or surprise disasters not only by promoting a proper understanding of the nature of "risks" in their community but also by strengthening their preparedness to "risks" in the whole cycle of disaster prevention, from the normal, emergency, and recovery phases .

Japan's National Research Institute for Earth Science and Disaster Prevention (NIED) has launched a five-year research project (2001–2005) with the aim of making modern societies resilient, not only to a traditional flood but also to possible catastrophic disasters of low probability but high consequence (LPHC). The approach that NIED has tried to explore is necessarily a new type of integrated risk management that should be tailored to the emerging disaster risks of the LPHC type. This includes (1) shifting the management strategy from "disaster prevention with zero risk" to "disaster reduction with an acceptable risk", (2) integrating both structure and non-structure measures (hard and soft measures), and (3) creating a social platform to call for a wide range of stakeholders (governments, communities, residents, corporations, local groups, and NPOs) in planning, designing and implementing an integrated risk management plan resilient to LPHC disasters in both short- and long-term perspectives.

In order to tackle these urgent tasks, a new team was formed on the basis of social and human sciences by adding fellows and guest researchers, unlike the conventional team of disaster scientists at the NIED. Then, the team began operations through close cooperation with outside research institutions such as the Disaster Prevention Research Institute of Kyoto University. Our research group has been developing such a participatory platform of disaster risk communication, called the Participatory Flood Risk Communication Support System (Pafrics), that can facilitate community-based participation in planning, designing and implementing processes for a better integrated flood risk management. Pafrics has been particularly developed by taking a number of research outcomes based on the social scientific studies concerning local people's flood risk perception and disaster prevention activities in their community through questionnaire surveys conducted by NIED. Currently, we believe that Pafrics has been developed to such a sufficient level as to be released on the Web, but it is still being repeatedly modified through trials and experiences with new aspects and episodes in new areas.

This book intends to provide outcomes of those studies in three parts. The first part expresses some of the important conceptual and methodological issues associated with the "integrated approach of disaster risk management" toward resilient society to emerging disaster risks in mega-cities in Japan. Four types of "integration" are taken into consideration in most of the papers. They are (1) integration of hard (structural control facilities) and soft (institutions and information) measures for shifting the concept of "zero risk" to "an acceptable level of risk", (2) integration of precaution and emergency relief measures, (3) integration of governmental and local community activities toward residents' informed choice of disaster risks, and (4) integration of individual risk management programs towards handling multiple sources of hazards. All papers focus on both the natural and socio-cultural factors of integrated risk management and their uncertainties involved in our modern society, reflecting the recent inter disciplinary development in "risk analysis", "disaster sciences", "resource economics", and "public policy analysis".

In the second part, all contributed papers are more or less associated with outcomes of case studies or the social surveys which NIED conducted since the start of the project. Those involved include the 2000 Tokai flood disaster, 2004 Fukui, Niigata Typhoon 23 disasters in Japan, as well as the 2005 Hurricane Catrena in the USA. Many important research topics are addressed in terms of public understanding or perception of disaster risks, public preparedness for reducing risks, attitude of local government officers engaging in disaster prevention, the role of volunteers in disaster prevention or relief activities, based on social scientific disciplines in relation to social psychology, disaster sociology, and disaster insurance and economics, etc. Those contributions are particularly important for the NIED project to look forward to making modern societies resilient to disasters of LPHC type by facilitating residents' participation to risk governance in local communities towards the informed choice of disaster risks.

The final part presents a set of papers which illustrate the development of "Pafrics", and some of the lessons we learned from several trials of using Pafrics in workshops, meetings, and lectures. In order to disseminate our model of "Pafrics", several internet-accessible versions are already open on the Web (http://www.pafrics.org), which is in Japanese for local residents, but a short English version is also available at the same Web site.

Finally, we should stress again that all contributions in this book are, more or less, the outcome of joint efforts conducted by all members of the project. However, final responsibility of the views expressed in these chapters lies with the individual authors themselves. We are very grateful for a number of invaluable comments and suggestions provided by the members of advisory body to our project; Dr. Sachio Kubo (Pasco Corporation), Mr. Nobuyuki Kurita (NPO: Rescue Stock Yard), Dr. Kimio Meguro (Tokyo University), Mr. Yukiji Nishida (NPO: Rescue Stock Yard), Dr. Norio Okada (Kyoto University), Dr. Yugo Ono (Hokkaido University), Dr. Isao Takagi (Keio University), and Dr. Kentaro Yoshida (Tsukuba University). We also express our deep gratitude for Ms. Reiko Shibakami, Ms. Reiko Kawamura, and other assistants for their sincere and endless effort to support our studies and preparation of these manuscripts during period of the project.

We should be very pleased if this book could make a valuable contribution towards a new perspective of "integrated disaster risk management" and "disaster risk governance" in the future.

March, 2006

Saburo Ikeda
Teruki Fukuzono
Teruko Sato
National Research Institute for Earth Science and Disaster Prevention (NIED)

Contributors

FUKUZONO Teruki
Storm, Flood and Landslide Research Department,
National Research Institute for Earth Science and Disaster Prevention
Landslide disaster

IKEDA Saburo, Dr. in Engineering
Emeritus Professor, University of Tsukuba
Visiting Researcher of NIED
Risk analysis, Systems science

MOTOYOSHI Tadahiro, Ph.D. in Educational Psychology
Department of Psychology and Human Developmental Sciences,
Graduate School of Education and Human Development,
Nagoya University
Educational psychology, Social psychology

NAGASAKA Toshinari, Policy Science, M.A.
Disaster Prevention System Research Center,
National Research Institute for Earth Science and Disaster Prevention
Risk governance, Risk communication

SATO Teruko, Geographer, M.A.
Head, Disaster Research Information Laboratory,
National Research Institute for Earth Science and Disaster Prevention
Flood disaster, Disaster risk management

SEO Kami, Ph.D. in Socio-Economic Planning
Aoyama Gakuin University
Risk analysis, Environmental economics

SHIMOKAWA Shinya, Ph.D. in Earth and Planetary Physics
Storm, Flood and Landslide Research Department,
National Research Institute for Earth Science and Disaster Prevention
Thermodynamics of the fluid system including ocean and atmosphere, and stochastic aspects of hydrologic and hydraulic phenomena

SUZUKI Isamu, Ph.D. in Human Sciences
Graduate School of Human Sciences, Osaka University
Volunteer studies, Group dynamics

TAKAO Kenji, Ph.D. in Social Psychology
Department of Clinical Psychology, Faculty of Health and Welfare,
Kawasaki University of Medical Welfare
Social psychology

TAKEUCHI Yukiko, Ph.D. in Science
Disaster Prevention Research Institute, Kyoto University
Geography, Education of disaster prevention science

TERUMOTO Kiyomine, Dr. of Urban Science
Disaster Reduction and Human Renovation Institution
Risk management, Urban and regional planning

TSUBOKWAWA Hiroaki
Visiting Researcher of NIED
Adjunct Lecturer of Kyoto University
Insurance Financial Engineering

ZHAI Guofang, Ph.D. in Socio-Economic Planning
National Institute for Land and Infrastructure Management,
Ministry of Land, Infrastructure and Transport
Urban and regional planning, Risk analysis, Environmental economics

Part I: An integrated framework of disaster risk management

A better integrated management of disaster risks: Toward resilient society to emerging disaster risks in mega-cities, Eds., S. Ikeda, T. Fukuzono, and T. Sato, pp. 1–21.

An Integrated Risk Analysis Framework for Emerging Disaster Risks: Toward a better risk management of flood disaster in urban communities

Saburo Ikeda

1 Introduction

For many years in Japanese urban communities, it has prevailed that the various flood prevention measures implemented upon major Japanese floodplains would prevent almost all large-scale flood damage to homes and buildings, even in flood-prone areas until we had the Tokai Flood of September 2000 in Nagoya. This common belief in the public was founded on the huge amount of public investment spent for construction of disaster prevention structural facilities after the Ise Bay typhoon caused 5,000 casualties in 1959. However, as the inconceivable damage, beyond our imagination, from the Tokai Floods of 2000 showed, the risk of catastrophic flood disasters has not been eliminated and should have continued to be considered by society. We have learned that we are seriously at risk from events for which we are unprepared, not because we fail to remember what happened before, but because we will encounter newly emerging risks that differ completely from what we faced before.

This is called "systemic risk" in a modern post-industrial society where a single physical disaster can trigger a spread of secondary and tertiary effects on other social systems or organizations, resulting in the collapse of entire systems supporting our economy as well as our social welfare (OECD, 2003). In such cases, many people could suffer long-lasting damage to their health as well as their property and environmental assets, due to severe impairment to urban and industrial infrastructure functions—such as communication, transportation, public health and security, disposal of sewage and industrial waste (e.g., hazardous chemicals, heavy metals). In addition to such systemic disaster risks, we face other risk issues arising from the construction of large-scale structural facilities to prevent disasters. Firstly, flood control projects, where the priority has been on early completion with economic efficiency, have degraded in the long-run river environments by reducing biodiversity, shrinking the habitats of aquatic fauna and flora, degrading the water environment, and changing the water-soil cycle. Secondly, rapid urbanization in the former flood-plain has weakened disaster preparedness on the part of local residents due to the decline of traditional local communities.

This paper is primarily concerned with the integrated risk analysis framework for analyzing the vulnerability of modern urban communities to the emerging disaster risks such as LPHC (low probability but high consequence) or multiple disasters in specific urban communities that have been technologically and socially constructed in modern society. Much attention should be directed toward community-based risk governance to cope with LPHC type of emerging disasters, particularly, in participatory and precautionary ways. The participation of local people in the planning, design, and implementation of flood-risk management measures could significantly reduce the scale of possible damage from catastrophic events of LPHC types.

Specific characteristics of our new framework are: 1) hazard research with interdisciplinary work for risk scenario formulation in terms of hazard sciences and human-social sciences, 2) risk assessment of emerging disasters under the high complexity and uncertainty involved in both human-socio-economic and natural-environmental systems, 3) integrated risk management under a wide range of value stakes among residents, local groups, NPOs and public administrative institutions by exploring various ways of integration regarding risk reduction measures in terms of "structural versus non-structural", "regulatory versus non-regulatory", "public versus market", and "single versus multi-hazards", respectively, and 4) disaster risk communication enhanced by a community-based social platform which supports stakeholders sharing previous disaster experiences of residents and hazard and risk information in the concerned area. Public administrative information about disaster management and the scientific knowledge of experts can also be provided through the platform. The residents, NPOs, public administration, experts, and other stakeholders can discuss mutual cooperation with nearby residents, and in the case of a disaster, find an ideal way to provide relief via the social network on the platform. Such a social system, which improves the level of disaster prevention through risk communication, is called "risk governance".

2 Integrated Risk Analysis Framework for Emerging Disaster Risks

2.1 Interdisciplinary concept of emerging disaster risks

It is not too much to say that many emerging natural disasters have been technologically and socio-culturally induced, because they are deeply associated with modern human settlement, production, transportation, and consumption activities in a post-industrial society. In addition, these risks are inevitably surprising and specific in their characteristics, typically of "low frequency but of catastrophic consequence" in terms of human life and health, economic loss, and associated environmental damage. Hence, such hazards are often comprehended as a possible danger as an event of "virtual-reality" in terms of the difference between the perceived possibility and reality, for example, the danger of encountering lightning or a airplane crash that are likely to occur no more than once in one's lifetime (Renn, 1992).

To deal with these types of emerging dangers, the modern concept of "risk" has evolved from the conventional view of "risk" as "an expected value of the probability of a hazardous event occurring times the magnitude of the consequence of the hazard" to an ontological or sociological concept of "risk" that allows us to take into consideration a wide range of socio-cultural characteristics of disaster risks. From among such attempts to define an interdisciplinary concept of risk, we take one of the simplest ones:

> "A potential for the realization of unwanted, adverse consequences to human life, health, property, or the environment."

This simple concept originated from extended discussions to define an interdisciplinary concept of risk at the Society for Risk Analysis (SRA: a professional and academic association founded in 1982 with major memberships in the US, Europe, and Japan). According to this concept, when we specify "a potential" in the framework of risk analysis, it should be noted that something reflecting human values as to "adverse or unwanted" has been included in the stakes considered through assessment and management processes of the experts, assessors, managers, and other stakeholders concerned with each risk event. In response to this issue of "values", either explicitly or implicitly inherent in any concept of risk, Kaplan and Garrick (1981) proposed the following expression of the risk concept as a risk triplet. The risk triplet consists of scenario S, likelihood P, and possible consequence D in relation to three basic questions of risk analysis:

1) What is the nature of disaster events that can occur?
2) How likely is a particular event?
3) What are the consequences?

$$\text{Risk} = R\{\langle S_i, P_i, D_i \rangle\}, \tag{1}$$

S_i: a set of scenarios concerning the nature of possible events

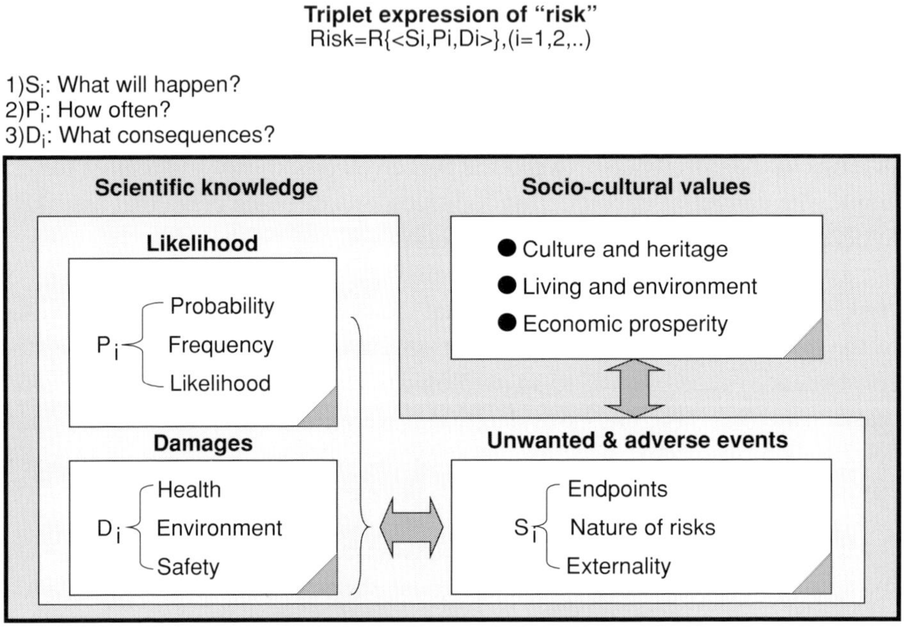

Fig. 1. The conceptual expression of "risk triplet" (Ikeda, 2000).

P_i: a set of likelihoods concerning event frequency, probability, or ambiguity

D_i: a set of consequences concerning potential damage to humans, animals, plants, and the environment.

The relationship between the conventional concept of risk $R\{\langle P_i \times D_i \rangle\}$ and the interdisciplinary one $R\{\langle S_i, P_i, D_i \rangle\}$ is illustrated in Fig. 1 (Ikeda, 2000). In addition to the scientific knowledge regarding P_i and D_i, scenario S_i can incorporate questions and conditions related to ontological and socio-cultural factors concerning our anthropocentric activities in a complex and uncertain world. For example, in the case of risk analysis for flooding events, the scenarios will try to identify what are the most critical endpoints to be assessed in terms of the possible impacts on humans, communities, and the environment to enable better risk management. As generic endpoint indices, we may argue in favor of the number of human lives lost, the economic loss with respect to household and community assets, damage to landscapes and environmental assets, and so on, depending on our concerns with respect to the nature of hazards.

Once we have suitable endpoints to be assessed, we have to specify the particular nature of each hazard that could affect the endpoints in both terms of the likelihood and the degree of damage. We also have to explore possible external impacts, which might occur accidentally or spontaneously in areas beyond our market or social institutions, by posing the following questions:

- Are the impacts or damage within our management limit or outside of our control?
- On what scale will these impacts be felt and what is their degree of irreversibility?
- Are these impacts inter-regional or transboundary in character?
- Do these impacts pose a potential threat to human ethics or morality?

These are typical questions that must be answered to clarify scenarios S_i in the problem formulation stage when we begin the scientific or objective evaluation of P_i and D_i because we may need other assessment schemes or tools tailored to the nature of each possible risk scenario. An important role of the risk analysis is to provide answers to these questions, either qualitatively or quantitatively, in relation to specifying the risk triplet $R = R\{\langle S_i, P_i, D_i \rangle\}$ based on an analytical framework of risk analysis described in the following section.

2.2 Typical risk scenarios of emerging flooding disasters in Japan

Figure 2 displays the salient issues and factors for generating risk scenarios that take into account the historical and cultural backgrounds in the context of hazards, river environments, and communities as discussed in details by other chapter (Sato). In this way, we can elaborate upon, for example, typical risk scenarios that require an integrated perspective in selecting adequate combinations of risk reduction measures:

Risk scenario 1: Increased flood-damage potential due to urban development and agglomeration

The accelerated economic growth and innovation in all social infrastructure sectors during the period from the 1960s through the 1980s brought about a rapid increase in the population of Japan's metropolitan areas. During this time, local rice paddies that once acted as a flood-protection belt were developed into residential and industrial districts. Such areas were vulnerable to

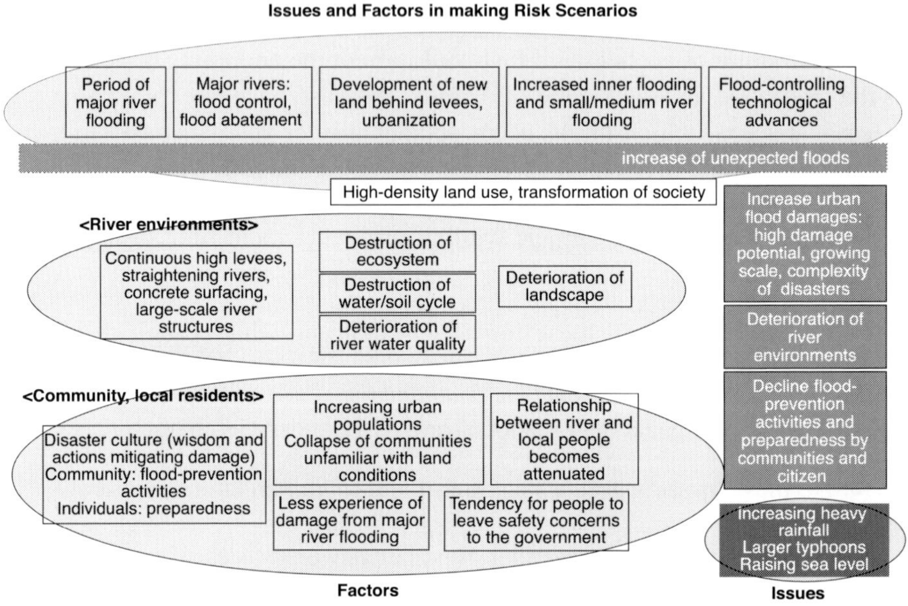

Fig. 2. Issues and factors in making "risk scenarios" for flood-disaster risk analysis (Sato, 2002b).

floods and frequently suffered damage as medium-sized and small rivers overflowed and inner flooding occurred. This problem leads to the rapid construction of rainwater drainage canals and flood-control facilities for such rivers. Flood damage in urban areas has declined, as indicated by a decrease in the total inundated area, with the construction of flood-control facilities. No decrease can be seen, however, in property damage due to floods over this 40-year period (Zhai and Sato, 2002).

Risk scenario 2: Increased systemic risk due to cascaded facilities or organizations

A number of small-scale problems in urban areas might lead to a series of large-scale hazards, increasing the systemic risk in a cascading manner, under specific conditions. In the Tokai floods, the inner drainage system failed to function properly for the first time during extraordinary precipitation—at a rate of 93 mm/hour, which exceeded the 100-year probability—that was nearly twice the capacity of the rainwater drainage canals in those areas. The overflow of medium-sized and small rivers continued until dikes, which were under the management of the local government failed at the Shin River. The area inundated by water was only 15% of the total inundation area, but accounted for 56% of the total property damage. A dreadful catastrophe would have occurred if the dikes holding back the largest river, which were under the management of the central government, had consequently failed (Sato, 2002a).

Risk scenario 3: Decline in preparedness of local communities to fight disasters

Urbanization has meant an increase in the number of new residents who are unfamiliar with the vulnerability of the local land to flooding. When danger from flooding is evitable, the government evacuates residents via a public warning system. This approach ensures an appropriate level of safety, but also reduces disaster-prevention awareness among the populace. It has become increasingly difficult for people themselves to prepare for an out-of-the-ordinary risk and to decide how to best protect their lives and property at the time of a disaster. In short, the trend is to "leave disaster prevention to authorities or other organizations", and residents wait for information from the authorities before acting when a disaster occurs (Sato *et al*, 2001). As a result, the ability of local communities to minimize losses from a disaster has been weakened. At the same time, a new movement has appeared in the form of disaster-prevention volunteers and non-profit and non-governmental organizations that provide assistance during disasters and deal with community safety of their own volition.

Risk scenario 4: Possible outbreak of LPHC type flood disasters

Continual development of man-made embankments along rivers has changed runoff characteristics and significantly increased the magnitude of flood discharge. Furthermore, embankment height has also significantly increased. As compared to 100 years ago, former excessive flood peak discharge now flows down river channels. This has enlarged the magnitude of flooding by potential failure of embankments, although, the probability of this occurrence is extremely low. In the 2004 severe flood disaster in Niigata Prefecture, the volume of floodwater that overflowed river channels, because of the failure of embankments of the Ikarashi and Kariyata Rivers, was assumed to be 40 to 50 times larger than the volume of flood water that overflowed the embankments. The large height difference between the top of the embankment and the ground made the destructive power of the flood-water strong. Once embankment failure had occurred in an urban area, tremendous damage would follow (Sato *et al.*, 2006).

Risk scenario 5: Environmental devastation and coastal erosions by interrupting soil-cycle

Artificial river channels have been increasingly constructed by flood control work for over 100 years since the introduction of former river law (1897), owing to strong public requirements for flood prevention. Fewer considerations were given to other functions of rivers than flood control. The river environment and wildlife habitat have been adversely affected as a result of public works. Prevention of downstream sediment transport, owing to the construction of dams for flood control and collection of large quantities of gravel in the high economic growth period, has caused riverbed degradation since the early 1960s. This degradation has necessitated further reinforcement of river structures, repairment of intake-weirs, and more restoration work of beach erosion as well (Sato, 2002b).

2.3 Conceptual risk analysis framework for natural disasters

As far as the generic issues of risk analysis posed in the risk triplet are concerned, our scheme of a risk-analysis framework can follow the classic one developed in the US by the National Research Council (NRC, 1983), which was primarily intended for the regulation of hazardous chemicals to protect human health and prevent environmental degradation. Although each step of a risk analysis may vary depending on the particular sources of the hazards, including both natural and man-made disasters, this has been widely accepted as a generic framework for regulating various types of risk problems. The framework consists of the three processes as shown in the upper part of Fig. 3:

1) Research processes (or problem formulation) to examine potential risk events, performed in laboratories, field studies, and communities, and in the case of disaster events, mostly based on scientific disciplines such as seismology, hydrology, climatology, environmental science, civil engineering, and so on.

2) Risk assessment, which provides an objective and integrated judgment in terms of scientific evaluations regarding hazard identification, pathways, and exposure assessment, and exposure-damage response assessment as an integrated form of risk characterization.

3) Risk management, a subjective decision-making process to select regulatory measures from among alternative regulatory options in conjunction with the outputs of scientific risk assessment and other socio-economic and cultural conditions.

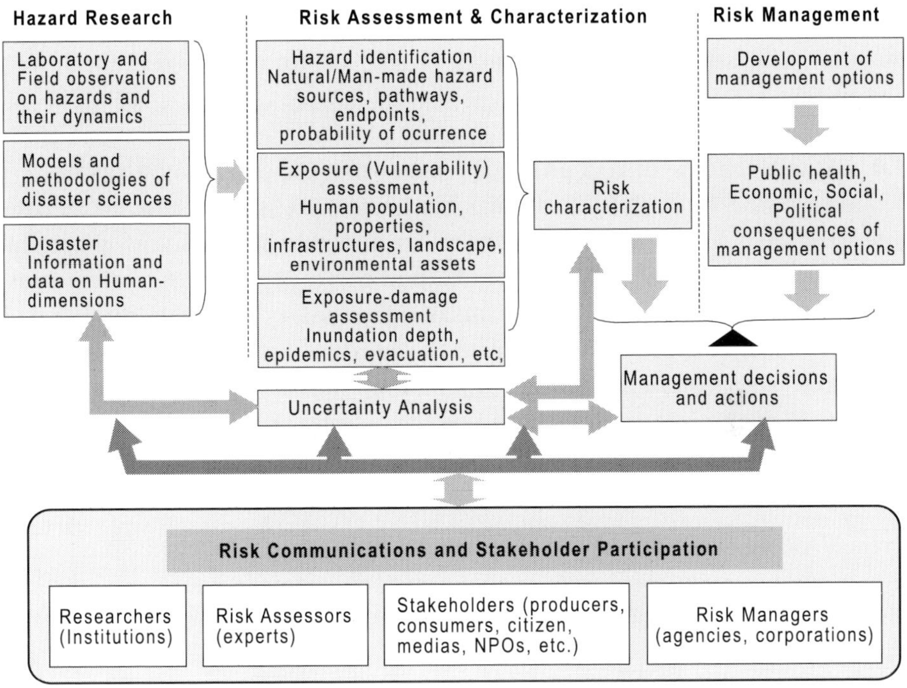

Fig. 3. Risk Analysis Framework for Natural Disasters (Modified from NRC, 1983).

One of the most important features of the NRC framework is the clear conceptual and functional separation between risk assessment and risk management. Although each process has its own tasks of risk characterization and management decision, respectively, both must deal with scientific uncertainty and complex stakes that are involved not only in its own process but also in the other process. The explicit representation of risk scenarios can provide an intermediate step or catalyst at the interface between these two types of processes to enable better assessment and regulatory decisions under high scientific uncertainty and complex stakes among researchers, assessors, managers, and stakeholders. However, we have seen many controversial cases of risk management which lost public acceptability or credibility primarily due to the lack of coordination regarding how to handle the uncertainty and complex stake issues between risk assessment and management processes. Typical examples were, in Japan, "Minamata disease" of organic mercury or "Itai-itai disease" of cadmium exposed through the complex food chain in the water environment. Hence, the critical role of risk communication and stakeholder participation in risk analysis has been widely recognized as a necessary fourth process in the risk analysis framework (NRC, 1996), as shown in the lower part of Fig. 3.

2.4 A new scheme of risk analysis framework for emerging disasters

In response to the NRC risk analysis framework, several models of disaster risk analysis framework have been developed to deal with the uncertain and complex nature of natural hazards, and these explore the relationships between hazards and damages in terms of either risk or vulnerability. For example, Wisner *et al.* (2004) advocated the pressure and release model (PAR model) as a means to understand risk in more realistic terms of vulnerability in the field of disaster sciences based on the disaster risk model of Alexander (1993)

$$\text{Risk} = \text{Hazard} \times \text{Vulnerability}. \tag{2}$$

In the PAR model, vulnerability is defined as "the degree to which someone's life, properties and other assets are put at risk by events in nature and in society", which can be measured in terms of potential damage.

In reference to the conventional expression of disaster risk provided by Smith (2001),

$$\text{Risk} = \text{Hazard(Probability)} \times \text{Loss(Damage)}/\text{Preparedness(Resilience)}, \tag{3}$$

Wisner's definition of "risk" as "hazard times vulnerability" can be interpreted as a unit measure of the possible damages D under an exposure probability P to the hazardous event in our risk triplet expression $R\{\langle S_i, P_i, D_i\rangle\}$ (1), where risk scenario S is implicitly taken into consideration as a degree of vulnerability in (2), or the preparedness in Smith's definition of disaster risk in (3). It should be noted here that our notation of "resilience" has just an inverse meaning of "vulnerability" to disasters. Rather, it has a more practical sense of enhancing "societal preparedness" to respond the emerging disaster risks.

Based on these concepts and measures of disaster risk, we can reach an integrated framework for disaster risk analysis which consists of the following four steps corresponding to the four parts of the previous NRC Framework (Fig. 3):

1) Hazard research (scenario formulation)—collection and analysis of data related to hazards in terms of their possible origins, pathways, and past

and present mitigation actions taken; based on this research process, risk scenarios can be developed.

2) Disaster risk assessment—preparation of a list of potential hazards along with their likely exposure or vulnerability, followed by risk characterization in terms of "risk curves" by both occurrence probability and damage outcome.
3) Disaster risk management—development of mitigation measures and procedures primarily based on the output from the risk characterization, taking into account uncertainty and other socio-human dimensions.
4) Disaster risk communication—creation of a platform to enable stakeholder participation in all processes of the risk-analysis cycle to help stakeholders understand the rationale behind risk assessment results and management options so that they can make better informed choices in uncertain and complex situations.

As a core part of our risk analysis framework, the process of disaster-risk assessment is schematically illustrated in the lower part of Fig. 4, which consists of the following three components and one integrative part of "risk characterization", given the possible risk scenarios in relation to the resilience or vulnerability in human/socio-economic dimensions and the potential sources of hazards in natural-environmental dimensions:

1) Identification of potential hazards of external forces either in natural-environmental dimensions or human-socio-economic dimensions. The likelihood of the strength of causal links between hazards and the endpoints (such as human casualties, damage to property, or impact on critical infrastructure) is also a critical issue to be taken into considered;
2) Assessment of the exposure to the hazardous events in relation to the resilience or vulnerability of human-dimensional systems, such as population, houses, assets, landscapes, etc.; and
3) Exposure-damage response such as the risk functions or curves between the degree of the exposure and damages based on possible mitigation measures;
4) Risk characterization will evaluate all issues of uncertainty and complexity involved in the preceding three steps, and will judge them in some integrated ways under a given acceptable or tolerable level of risk, which presumably should be considered as historically acceptable or being implicitly rooted in the concerned society with regard for safety from natural disasters.

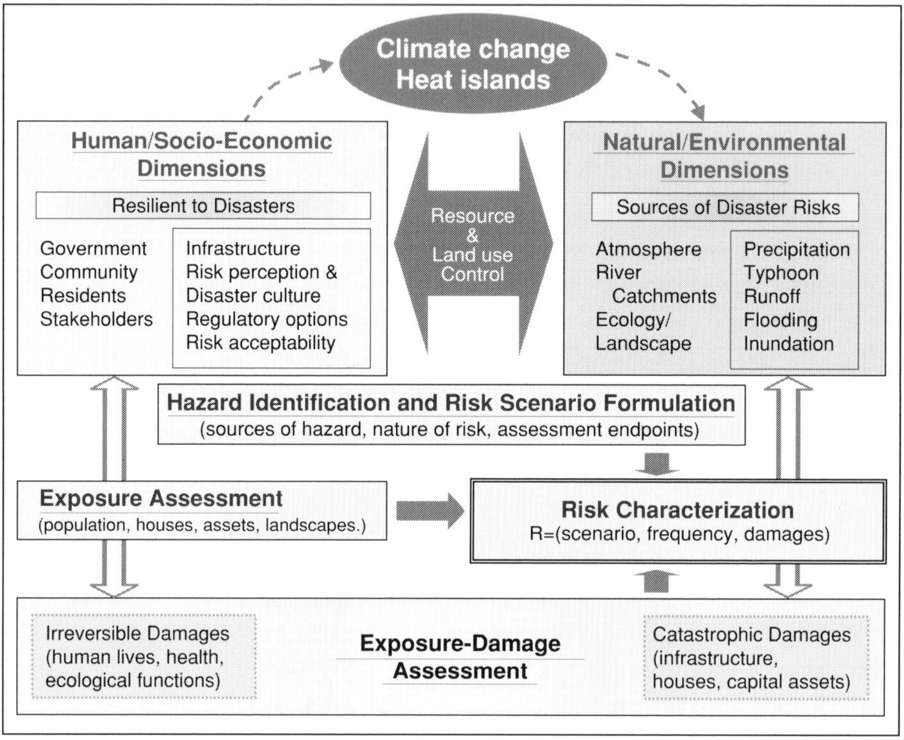

Fig. 4. Schematic flow of risk characterization in disaster-risk assessment.

3 Risk Management Strategies for Better Governance to Emerging Disasters

To figure out the salient issues in disaster-risk management strategies for better governance, we can classify the nature of risks into four categories by dividing them according to two perpendicular axes, with one axis indicating the degree of uncertainty of our knowledge (a scientific evaluation axis related to a causal structure) and the other axis indicating the degree of social stakes among the stakeholders involved in the risk events (a socio-economic or cultural and ethical evaluation axis associated with a disutility structure) (Ikeda, 2000).

Area 1: This is an area where scientific knowledge about risks is fairly certain and it is little stake among stakeholders for evaluating the results of the scientific assessment using objective indices or standards. Here, we can easily implement regulatory measures to reduce risks based on objective risk assess-

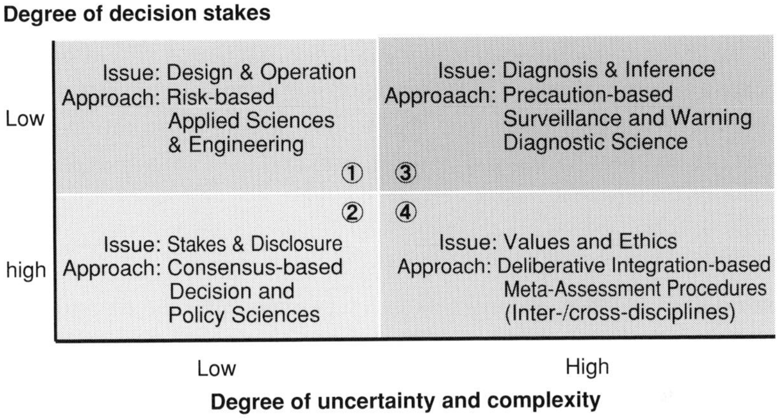

Fig. 5. Risk management issues and approaches in the uncertain and complex world (Ikeda, 1996).

ment. The important management issue is how to ensure that risk information is objective in maintaining the accountability of regulatory decisions, selected from among possible options, to reduce risks to a level lower than acceptable levels. Hence, the appropriate management strategy is to take a risk-based regulatory approach under the existing legislative frameworks and institutional settings.

Area 2: This is an area where scientific knowledge about risk events is fairly certain, but there is disarray among stakeholders regarding their evaluation of the risk-assessment outcomes. In this area, a consensus-building approach to risk management is required as the principal management strategy. The critical issue is to ensure that the risk assessment procedure is reliable and transparent, not only by having stakeholders involved at all levels of regulatory decision-making, but also by making democratic institutional arrangements to ensure their involvement.

Area 3: This is an area where the scientific knowledge is uncertain, but stakeholders feel that they have less at stake in evaluating the risk assessment results, even in an uncertain and ambiguous context. Some risk events corresponding to natural or environmental hazards, such as earthquakes or volcano eruption, would be located in this area. Here, the risk assessment must take the form of a diagnosis, prediction, or scenario in terms of qualitative or subjective measures under a high level of uncertainty. The main management strategy is how to promote "sound science of diagnosis and presumption" despite a high level of uncertainty in our scientific knowledge. Hence, it would be desirable to allocate a significant share of resources not only to surveillance systems for detecting early warning signs but also to developing "precaution-based risk communication" for sharing risk information and data among experts and the public.

Area 4: This is an area where the scientific knowledge is uncertain and significant conflict is likely among stakeholders evaluating the diagnosis or presumptions in relation to an acceptable or tolerable level of risk. Since several existing approaches based on scientific and objective indices or standards are almost impossible to employ, it is necessary to develop a meta-science of deliberative integration approach to evaluate both scientific and socio-cultural factors associated with human dimensions of risk problems that include anthropological, ethical, and value judgments. At the same time, as in Area 3, we need to enhance "risk communication" platform for the stakeholders to foster their integrated perspectives of risk governnance in relation to both human and natural dimensional issues.

In Areas 1 and 2, we can scientifically evaluate possible events qualitatively or quantitatively with a reasonable degree of certainty, such as the probability of precipitation being greater than 100 mm per day or the odds ratio of excess economic damage caused by floods based on measured inundation depth. The important management issues then become 1) how to ensure risk scenarios and the related indices or risk measures are properly chosen, monitored, and regulated, and 2) how to achieve reliable decision-making or consensus-building through the participation of stakeholders in these processes. Most disaster risk management, except that concerning surprising or catastrophic events, falls into these domains provided we have fairly good scientific monitoring data to allow objective assessment.

However, since most emerging risks correspond to surprising or catastrophic hazards, which inevitably are associated with high scientific uncertainty or ignorance, they generally fall into Area 3 or 4. Whether the main management issues are located in either Area 3 or 4 depends on the context of socio-cultural stakes among the interested groups or actors, whose risk perceptions will have different roots depending on their ontological or anthropological perspectives, which lie outside of conventional science paradigms. Here, we need a "societal risk governance" approach that is centered around the precautionary framework primarily supported by discursive-type risk communication among stakeholders (Renn and Klinke, 2001). We use "risk governance" to represent a new integrated type of risk management strategies in which interdependence or intercommunication among stakeholders is essential for them to pursue collective decisions and attain policy goals, particularly through socio-political-cultural networks both vertically and horizontally. It is also assisted by a variety of networks enhanced by social networking systems among volunteer groups and other stakeholders, which we call e-community, a virtual type of community implemented in the internet web by utilizing a variety of information technologies of web-log and web-GIS systems together with mobile technologies as described in other chapter (Nagasaka, 2006).

4 A Pilot Study of Risk Governance to Emerging Flood Disasters in Urban Communities

4.1 A NIED project

Japan's National Research Institute for Earth Science and Disaster Prevention (NIED) has launched a five-year research project (2001–2005) with the aim of making modern societies resilient not only to a traditional natural disaster but also to emerging disasters such as LPHC (low probability but high consequence) or multi-disasters that have technologically and socially induced in modern society. The project put specific emphasis on the following policy issues which recent Japanese disaster management had began to look for:

1) Shifting the management strategy from disaster prevention with zero risk to disaster reduction with an acceptable level risk that the residents or local communities may take.

2) Integrating a variety of risk reduction options in terms of hardware versus software, precautionary versus recovery, or regulation versus market/volunteers for making better governance to emerging disaster risks.

3) Facilitating both residents and stakeholders (local regulatory authorities, communities, NGOs) participation in planning, designing, implementation, and monitoring processes.

In order to cope with such policy issues, we have been developing such a social platform of assisting disaster risk communication, called Participatory Flood Risk Communication Support System (Pafrics), that can facilitate community-based participation in planning, designing, implementation, and recovery processes. The Pafrics has been constructed by taking a number of research outcomes concerning local people's risk perception and disaster prevention activities through a series of questionnaire surveys conducted by NIED (Sato *et al.*, 2003).

Figure 6 shows our risk analysis procedure for dealing with emerging flood risks exposed to communities of highly urbanized areas. Although this new procedure of risk analysis follows the generic framework of disaster-risk analysis, as illustrated in Fig. 3, it has specific emphasis on the utilization of a social platform of a risk communication system in its management process as a social experiment of risk communication in local communities. The whole procedure consists of four major steps in which each task of the NIED project is allocated as written on the right-hand side in Fig. 6. In the first step, alternative risk scenarios are created as the problem formulation taking into consideration the human-dimensional and natural-environmental factors and the social issues displayed in Fig. 4. In the second and third steps, the risk is assessed either qualitatively or quantitatively, paying careful attention to the uncertainty involved in each process of hazard assessment, exposure assessment, and exposure-damage response, by forcusing on a combination of hard and soft measures to reduce the flood risk. In the fourth step, a new type of flood-risk communication support system, Pafrics, which involves resident participation, is developed and tested.

During period of the project development, socio-economic and psychological studies regarding the social vulnerability to emerging disasters, are being carried out to develop methods for raising disaster-prevention awareness in flood-prone areas or for creating a mechanism of informed choice by having stakeholders' participation in the collective decision-making of choosing alternative prevention measures including the study of risk finance or insurance institution. The seven research topics of the NIED project appearing in Fig. 6 are:

1) Structural and simulation analysis of disaster-risk occurrence, propagation, and associated damage based on risk analysis.

2) Risk-based assessment of disaster-prevention measures within the conventional scientific framework of probabilistic approach to natural disasters.

3) Integrated risk characterization including both human and environmental dimensions with specific emphasis on economical, cultural, and psychological considerations.

4) Exposure and vulnerability assessment together with community-based preparations for disasters mitigation in terms of precaution-based or deliberative integration-based management.

5) Development of a participatory flood risk communication support system (Pafrics) to help users to obtain a deeper understanding of the nature of flood risk and management options for their collective informed choice among stakeholders. (http://www.pafrics.org).

6) Risk finance and institution for flood disaster funds for the LPHC type of catastrophic disasters as one of the critical options.

7) Community-based integrated framework of multi-disaster risk governance based on the e-community enhanced by information technology.

Although each task is discussed in details in other chapters of this book, it should be stressed here that, while a construction of a social platform of risk communication support system (Pafrics) operates as a warp for the NIED project, the concept of integration in various measures of risk reduction works as a weft of all tasks in the NIED project. Here are listed such ways of integration, depending on the nature of a pertinent risk in terms of the degree of uncertainty in our knowledge and the degree of social stakes among stakeholders as illustrated in Fig. 5.

i) Integration of hard (structural control facilities) and soft (institutions and information) measures for shifting the concept from "disaster prevention with zero risk" to "disaster reduction with an acceptable or tolerable level of disaster risk":

Under the present conditions, much attention should be directed towards soft measures consisting of institutional arrangements such as land-use regulation, risk finance schemes for disaster insurance, dissemination of early alarms or risk information through hazard maps or media, and provision of economic incentives for public participation in emergency preparation or drills.

ii) Integration of risk-based and precaution-based measures or actions throughout the disaster risk management cycle:

It is always desirable to take balancing perspectives between scientific uncertainties and complex processes of cause and effect relationship such as long-term economic efficiency, political flexibility, sustainable development of communities, and preservation of the natural environment at each stage of normal, emergency, and recovery processes.

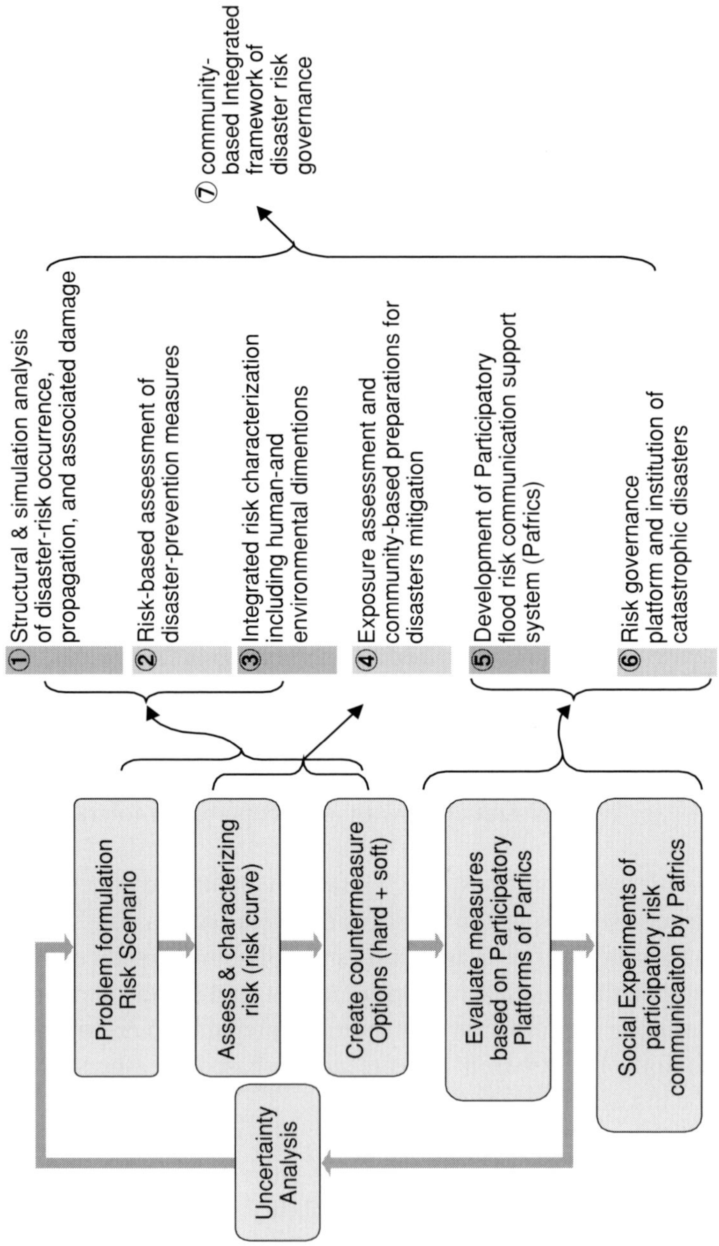

Fig. 6. Integration procedures for the NIED project.

iii) Integration of regulation, market, and community-based volunteer activities in planning, design, and monitoring activities by promoting the participation of local residents:

Since both the government and local communities are limited in terms of their budgets and human resources, it will be necessary to return to such basics of disaster prevention culture based on an "informed choice of risk". Here, individuals take on the responsibility of protecting their lives and property given the adequate provision of risk information either by governmental sectors or by the mutual support of the local community and relief provided by the social network of citizen volunteers.

iv) Integration of single risk management program toward multiple sources of disasters in local community:

There are various types of socially constructed disasters such as trafic accidents, fires, criminal activities, or terrorism and so on at the level of local communities. However, the social structure of most communities is rapidly becoming one of an aging society with less knowledge of ways to deal with these multi-disaster issues. In this situation, it is urgently requested that the local communities adopt positive attitudes in favor of activating the mutual support in their neighborhood and relief provided by the social network of citizen volunteers fostered through e-community platform.

5 Concluding Remarks

To have a better societal governance of the emerging disaster risks triggered by phenomena such as rapid urban agglomeration and complex processes of infrastructure development in our post-industrial society, we have developed an integrated risk analysis framework for finding ways of making modern urban society resilient to such disasters. Much of our attention has been directed towards community-based risk governance to enhance our preparedness for disaster risks by integrating a variety of risk-reduction options in terms of hardware versus software, precaution versus recovery, or regulation versus market. For making better governance to emerging disaster risk, it is critical to improve the mutual cooperability among residents and stakeholders (regulatory authorities, community, volunteer groups, NGOs) in planning, designing, implementation, and recovery processes of disaster governance.

We have described four ways of integrating risk-reduction measures, emphasizing the soft measures of implementing institutional arrangements such as land-use regulation, risk finance schemes to provide disaster insurance, dissemination of early alarms or risk information through hazard maps or media, provision of economic incentives to encourage public participation in emergency preparation, and so on. Our focus is partly based on the fact that, as concern grows regarding the emerging disaster risks of increased potential for catastrophic disasters in our mega-cities, we face an urgent need for an integrated approach to the systemic risk in which human-environmental factors will be critical to the societal governance of emerging disaster risks.

Finally, we should point out that alluvial lowlands—where Japan's mega-cities are located and which are especially vulnerable to flood disasters—are essentially new lands formed within the last 10,000 years. This means that these urban areas are eventually vulnerable to earthquakes and other natural or man-made disasters. When discussing the means of reducing disaster-related damage in Japan's mega-cities, a comprehensive and integrated approach to multi-hazards must be taken as opposed to treating earthquakes, floods, and fires separately. This is among the next tasks we will explore.

References

Alexander, D. (1993): *Natural Disaster*, UCL Press.

Ikeda, S. (2000): Strategy, policy and institutions for risk management, in *Handbook of Risk Research* (ed. Society for Risk Analysis Japan-Section), pp. 306–311, TBS Britannica (in Japanese).

Ikeda, S. (2004): Risk society and planning approach toward social governance on technological and environmental risks, *Planning and Administration*, **27**(3), 3–11 (in Japanese).

Ikeda, S., T. Sato, and T. Fukuzono (2006): Integrated Framework for Disaster Risk Management: A NIED Project of Societal Governance for Emerging Disaster Risks, Proceedings of Fifth Annual IIASA-DPRI Forum on Integrated Disaster Risk Management, Beijing. (http://www.iiasa.ac.at/Research/RMS/dpri2005/proceedings.html).

International Risk Governance Council (IRGC) (2005): White Paper on Risk Governance, (http://www.irgc.org/_cgidata/mhscms/_images/13779-3-1.pdf) (by Feb. 10, 2006)

Japan Meteorological Agency (JMA) (2005): 2005 Report on climatic aberration, 382 (in Japanese).

Kaplan, S. and B. Garrick (1981): On the quantitative definition of risk, *Risk Analysis*, **1**(1), 11–28.

Ministry of Land, Infrastructure, and Transport (MLIT) (2003): "Flood Statistics, 2002", 515, (in Japanese).

Nagasaka, T. (2006): New mode of risk governance by e-community platform, Chapter I-5, *A Better Integrated Governance of Disaster Risks* (eds. Ikeda *et al.*), TERRAPUB, Japan.

National Research Council (1983): *Risk Assessment in the Federal Government*, National Academy Press.

National Research Council (1996): Understanding Risk: Informing Decisions in a Democratic Society, National Academy Press.

OECD (2003): Emerging systemic risksin the 21st Century: and agenda for action, Paris.

Renn, O. (1992); Concept of risk: A classification, in *Social Theories of Risk* (ed. Krimsky, S. and D. Golding), 53–79, Praeger Publisher.

Renn, O. and A. Klinke (2001): Environmental risks—perception, evaluation and management: epilogue, in *Environmental Risks: Perception, Evaluation and Management* (eds. Boehm, G. et al), 275–299, Elsevier Science.

Sato T. (2002a): Time and spatial distribution of damages in the 2000 Tokai urban flood disaster, Natural Disaster Report 38, NIED (in Japanese).

Sato, T. (2002b): "Some discussion on social systems to reduce flood risk", Geography, Kokon Shoin Pub. Co., Vol.47-2, 48–54 (in Japanese).

Sato, T. *et al.* (2001): "Residents' disaster perception and their preparedness: 2000 Tokai Flood Disaster case", Proceedings of Second Asian Symposium on Risk Assessment and Management, 40–50.

Sato, T., T. Fukuzono, S. Ikeda, and K. Seo (2003): Development of Participatory Flood Risk Communication Support System, Proc. of Annual Meeting of Japan Society for Natural Disaster Science, 223–224 (in Japanese).

Sato, T. *et al.* (2006): "Characteristics of the flood risk of the 2004-Niigata flood disasters", Natural Disaster Report 40, NIED (in Japanese).

Smith, K. (2001): Environmental Hazards: Assessing risk and reducing disaster (3rd edtion), Routledge UK.

Wisner, B., P. Blaikie, T. Cannon, and I. Davis (2004): *At Risk: Natural Hazards, People's Vulnerability and Disasters* (2nd edition), Routledge UK.

Zhai, G. and T. Sato (2002): Structural Analysis of 2000's Tokai-Flood in Japan: Towards Establishing a Framework for Megalopolitan Flood Risk Analysis. IIASA-DPRI2002 Proceedings: http://www.iiasa.ac.at/Research/RMS/dpri2002/Papers/Zhai.pdf

A better integrated management of disaster risks: Toward resilient society to emerging disaster risks in mega-cities, Eds., S. Ikeda, T. Fukuzono, and T. Sato, pp. 23–40.

Fundamental Characteristics of Flood Risk in Japan's Urban Areas

Teruko Sato

1 Introduction

Flood risk is diverse and complex. Risk-related phenomena such as flood-inducing precipitation, runoff generation and concentration, downstream flood wave propagation, flooding, and flood damage are changing over time and vary from region to region under the influence of natural conditions, human activities, and Japan's disaster culture. In Japan, the loss of life and national economic losses caused by flooding have drastically declined over the last 60 years. However, new flood risks are emerging in urban areas, including more potential for flooding, more exposure to flood risk, and new forms of damage.

For example, in Japan, recent flood disasters resulting from embankment failures have led to catastrophic damage and consequences. The general economic loss caused by the 2000 Tokai flood disaster, in which the metropolitan areas of Nagoya were flooded, was the worst in 40 years. Furthermore, embankment failures led to serious damage in a number of provincial urban areas in 2004, with the drowning of elderly people a major issue in Niigata. New risks have also been identified in urban areas. In 1999, the underground shopping mall at Hakata Station in Fukuoka, Japan's sixth largest city, was flooded, and one life was lost in a building basement. River environments have undergone drastic changes over the past 100 years as rivers are forced into artificial channels, leaving them with less natural, more artificial environments. Moreover, disaster-prevention awareness and activities in local communities has been decreasing.

Takahashi (1964, 1971) examined flooding events in Japan and showed that floods were not a purely natural phenomenon and that social conditions played an important role that varied from region to region and from time to time. Takahashi (1964) illustrated the structure of modern flood disasters up to the 1960s, just the initial stage of Japanese experience in a period of high economic growth. He also pointed out that the peak flood discharge of rivers had increased as a result of river improvement projects. Today, more than 40 years after the report, flood disasters have taken on a new look and flood risk has become more diverse and complex.

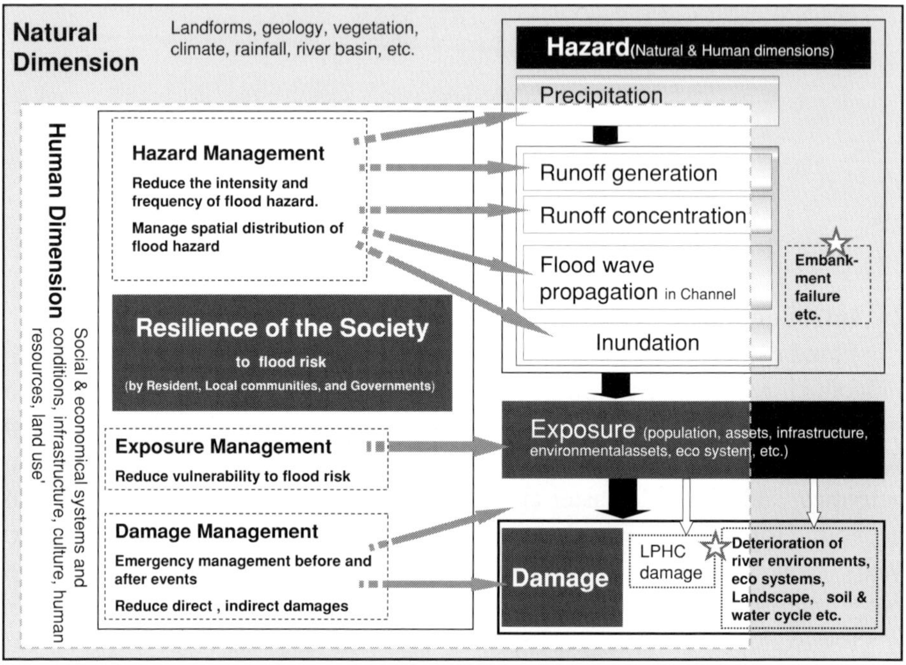

Fig. 1. Causal structure of flood risk.

This paper describes the recent structural changes in flood risk characteristics for urban areas in Japan using the results of flood-disaster investigations into the 2000 Tokai urban flood disaster which occurred in one of the metropolitan areas in Japan and the 2004 Niigata flood disaster which occurred in provincial urban areas (Sato, 2002; Sato *et al.*, 2006). The paper reviews existing knowledge in the fields of geography and river engineering and uses the causal structure of flood risk to describe the characteristics of flood risk.

2 Causal Structure of Flood Risk

The causal structure of flood risk is shown in Fig. 1. Flood risk has four components: Hazard, Exposure, Damage, and Social Resilience to flood risk. All four components are subject to both natural and artificial environments that vary over time as social and economic conditions and human activities evolved. The components are defined as follows;

1) Hazard is an external natural force that has the potential to cause flood damage. Precipitation is the primary external factor. The scale and pat-

tern of precipitation and its distribution in time and space in a basin are major factors that determine the magnitude and characteristics of the flood hazard. The flood hazard is transformed, as shown at the far right of Fig. 1, as precipitation runoff concentrates in river channels and flood waves propagate down the river channels. Ultimately, inundation occurs. Because these phenomena occur on or near the surface of the Earth, where human activities take place, flood hazards have local characteristics that are affected by geology, topography, vegetation, and land use in the river basin.

2) Exposure refers to the socio-cultural units that are exposed to flooding, such as populations, assets, environmental resources, biological resources, lifelines, social functions, etc.

3) Damage is the deterioration or loss of functionality of units exposed to flooding. This includes not only direct damage but also indirect damage to the social, economic, and natural environments.

4) Social Resilience to flood risks is the ability of society to withstand the hazard as a result of efforts in disaster risk management.

3 Characteristics of the Recent Flood Risk in Japan

3.1 Flood hazards

One of the characteristics of flood risk in Japan is the significant qualitative and quantitative changes in the flood hazard resulting from human activities such as flood control measures, a trend that grows stronger during periods of high economic growth. As a result, the frequency and intensity of floods has drastically decreased, increasing the possibility and complexity of catastrophic flooding and creating a cascading series of flood hazards in urban areas.

3.1.1 Outline of changes in flood hazards

Flooding can be effectively controlled using Large-Scale Flood Control Structures (LFCS) such as high continuous embankments along rivers, as well as dams, pumping stations, etc. These LFCS have been applied to a remarkable degree throughout the length and breadth of Japan for 100 years along both large and small rivers. As a result, the intensity and frequency of floods, in terms of spatial and temporal distributions, have drastically decreased in Japan over time. Alluvial lowlands are no longer constantly vulnerable to flooding.

However, by improving and shortening the river channels, constructing high and continuous embankments to keep flood waters in river channels, and expanding storm-water drainage systems, LFCS have increased a potential of creating new flood hazards. Namely, LFCS change the way flood waves propagate in rivers, shortening the time-lag between the rainfall and the peak discharge, thereby increasing the flood discharge flowing down the channels (Takahashi, 1971; Sato, 1998). Moreover, the volume of flood runoff has increased as a result of the loss of water detention capacity of urban catchments. These days, a particular quantity and pattern of rainfall results in a flood discharge of greater volume and with a higher peak discharge than ever before. Moreover, the number of heavy rainfalls, the primary external force of a flood hazard, has been increasing in urban areas in Japan, according to statistics for Japan (JMA, 2005). If this trend continues, it might become a major factor in increasing the flood risk in urban areas.

In addition, another factor that increases flood risk is occurring in the floodplains. This factor is ground subsidence and is mainly caused by groundwater pumping. This has increased the area, depth, and duration of floods. Sometimes subsidence has led to the sinking of river embankments.

3.1.2 Increasing the possibility of catastrophic flooding

The development of LFCS has notably decreased the frequency and intensity of flooding by cutting off the process of hazard transformation, as shown on the right-hand side of Fig. 1. However, recent flood disasters in Japan indicate a different scenario for catastrophic flooding: embankment failure resulting from precipitation exceeding the level specified in the design. In other words, extensive LFCS construction in Japan has increased the possibility of catastrophic flooding.

The typical process of how catastrophic flood hazards escalate is clear from the 2004 Niigata flood disasters. The Ikarashi River and Kariyata River are branches of the Shinano River, the longest river in Japan. The catchments are 239.8 km^2 for the Kariyata River and 310.1 km^2 for the Ikarashi River. The following is a description of the process by which the flood hazards escalated, leading to the 2004 Niigata flood disaster.

1) Escalation of the flood hazard potential as a result of LFCS construction

i) Increases in embankment heights and channel capacity The channel and embankments of the Ikarashi River have been repeatedly improved in response to severe floods. A great flood in 1873 triggered the start of river improvement works in 1876. In 1925, the other severe flooding resulted in additional work to increase the flood flow capacity of the channel, which reached 1,120 m^3/sec in 1937. The capacity was further increased to 1,600 m^3/sec after a flood in 1961. After yet another flood in 1969, the capacity was increased to 2,400 m^3/sec. In addition, an upstream dam increased the design flood flow of the channel to 3,600 m^3/sec. Thus, the river capacity has risen continually through channel and embankment work after major disasters. The peak flood discharge currently is 2.14 times larger than it was 70 years ago. In a similar manner, the flood capacity of the Kariyata River channel has risen from 650 m^3/sec in 1920 to 950 m^3/sec, 1,050 m^3/sec, and finally 1,550 m^3/sec in 1969. With this capacity, it is designed to accommodate the type of floods that occur only once in a century. The frequency and intensity of flooding has drastically decreased on the floodplains, but the embankments have risen in height and, with heavy precipitation, a huge discharge flows down the river channels (Sato, 2006). Thus, by repeatedly reducing flooding, this embankment work has ironically or necessarily increased the potential for catastrophic flooding of due to possible embankment failure.

ii) Embankment failures caused by rainfall exceeding design specification During the 2004 floods, heavy rainfall induced a peak flood discharge of 1,900–2,000 m^3/sec in the Ikarashi River and 1,700 m^3/sec in the Kariyata River, respectively, exceeding the design discharge of the river channel and causing embankment overflow. The embankment had been eroded and eventually collapsed. The level of precipitation contributing to the peak discharge of the two rivers occurs only once every 500 years (JSCE, 2004). For example, 267 mm of rainfall fell in six hours. In twenty-four hours, the amount reached 422 mm, far above the level specified in the design of the LFCS.

iii) Increased force and volume of floodwaters Floodwaters generally flow very slowly over an alluvial plain that has a very gentle slope of 0.7–0.8/10,000, as in the case of the floodplains of the Ikarashi River and Kariyata River. However, the embankment failure increased the force of the floodwaters because of the big difference in the hydraulic head between the level of water in the river channel (14.7 m) and that of the floodplain (9.46 m) and the very short collapse time (with 10 m of the embankment collapsing in just 5 min). In the end, the breach was 50 m wide (Niigata Prefecture, 2005). A huge and rapid flood flow washed away a temple near the collapse site, and houses within 150 m from the site were completely destroyed.

The estimated volume of floodwater released by embankment failures along the Ikarashi River was 13.93 million m^3. Along the Kariyata River, it was 8.80 million m^3. These volumes are 40 to 50 times larger then the estimated volumes of water that simply overflowed the embankments of these two rivers (0.34 million m^3 and 0.36 million m^3, respectively). The observed result in both cases was much deeper and more extensive flooding (Sato *et al.*,

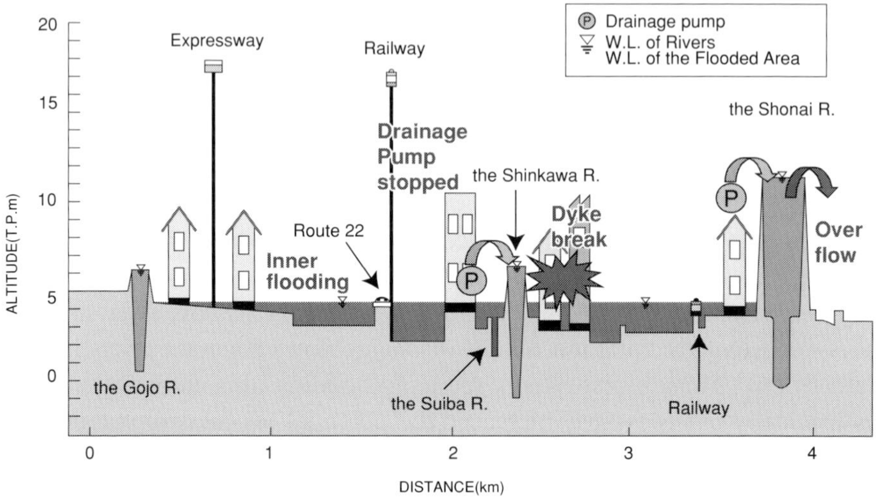

Fig. 2. Topographical cross section of the 2000 Tokai floods disaster area Increase of flood damages by different causes in the 2000 Tokai-Flood (MLIT, 2000).

2006).

2) Increased depth of flooding by development of floodplain

Another factor that contributed to the intensity of the flooding was development of the floodplain. One example is the development of the narrow valley plain on the left side of the Kariyata River in Mitsuke City. River improvement works had straightened the meandering channel but narrowed the width of the valley from 400–500 m to 300 m. Housing development in the narrow valley plain accompanied the work. The floodwaters flowing from the embankment breach site were dammed by houses, factories and earth mounds, which deepened the floodwaters, allowing them to reach older settlements on the higher terraces that used to be relatively safe from flooding (Sato *et al.*, 2006).

3.1.3 Increase in complexity, and cascade effect of flood hazards

Fig. 2 shows a typical river system in an urban area with considerably developed LFCS type river improvements. This example is a topographical cross-section of the 2000 Tokai flood disaster area, the Nagoya metropolitan area. Flooding caused by large rivers had been reduced. On the other hand, rapid urbanization led to increases in flooding by small and medium-size rivers. This led to various management bodies making improvements to urban rivers of various sizes.

Fig. 2 shows how rivers with different runoff characteristics form an interacting network, with sizes ranging from rainwater drainage systems to large-scale rivers. This means that a variety of hazards arise in a cascading manner in any one area, such as from inner flooding, flooding caused by small and medium-size rivers, and catastrophic flooding triggered by embankment failure on a large river. Furthermore, the water levels of rivers affect each other on an alluvial plain with a low gradient, strengthening the potential of the cascade effect. As shown in Fig. 2, if the Shonai River (a large river) or the Shin River (a medium-size river) becomes full, the drainage pumps must be halted, which intensifies the flooding on the floodplain. Water levels in tributaries such as the Gojo River and Suiba River (both small rivers) are affected by the backwaters from the Shin River. Moreover, there is no integrated management system among these rivers. Improvement plans for each are designed on a different scale, and they are managed by different management entities. For example, large rivers are managed by the central government, which plans for large floods that occur once in 200 years. Medium-sized rivers are managed by the prefectural office, which designs plans to handle floods that may occur once every ten to fifty years. Small rivers and urban storm drains are managed by the city office, which designs plans to handle floods that may occur once every ten years.

Heavy rainfall in excess of the levels specified in the design of the LFCS measures triggered the 2000 Tokai flood disaster. For example, the 97 mm of precipitation recorded in 60 min. at the Nagoya meteorological observatory occurs once every 110 years and exceeds the design specifications of the storm-water drain system and the small river improvement works. A 24-hour precipitation of 534.5 mm was recorded. Such precipitation occurs once every 350 years and exceeded the design specifications for improvement works on medium-size and large rivers.

Fig. 3 illustrates the cascade effect of flood hazards in the 2000 Tokai flood disaster areas. The inundated area and the area of economic loss increased as damage and inundation accumulated from different hazard sources during the event (Sato, 2002). First, inland flooding was caused by an overflowing storm-water drain system that had the smallest catchment. The flooding occurred on September 11 at around 18:00, immediately after 97 mm of precipitation in 60 minutes. The floodwaters rose over the top of the embankments of the Yamazaki River, a small river. The water level then rose to the high-water level specified in the design for medium-size rivers, at which point, pumping was stopped to prevent overflowing of the embankments. Then overflow caused the Shin River, a medium-size river, to fail its embankment, resulting in severe damage. Finally the floodwaters began to overflow at some part

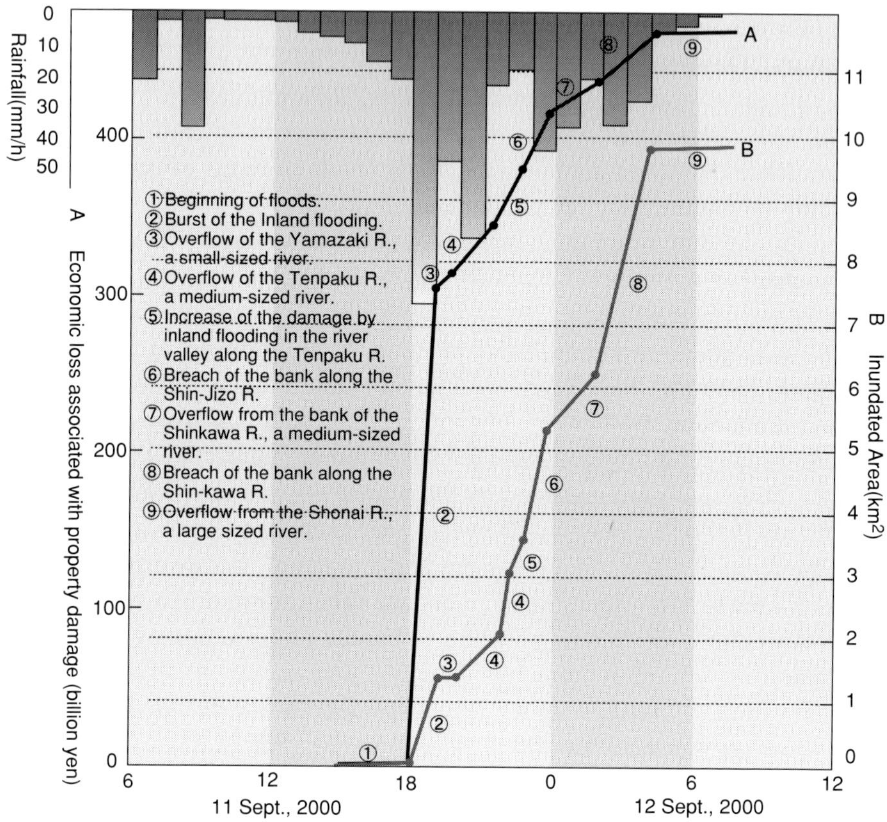

Fig. 3. Increase of the damage caused by flooding of different rivers in 2000 Tokai Flood (T. Sato, 2002).

of the embankments of the Shonai River, the largest river, on September 12 at around 4:30, approximately ten hours after the floodplain began flooding. Sandbagging prevented overflowing, and the embankment did not fail.

3.2 Change in exposure and new type of damage

The damage caused by flooding has drastically changed both quantitatively and qualitatively through urbanization and the more extensive land-use in urban areas. In these areas, the potential for damage has increased along with the possibility of catastrophic damage from flooding. On the other hand, a new type of flood damage has emerged.

3.2.1 Catastrophic damage caused by flooding

Accelerated economic growth and innovation in all sectors of the social infrastructure in the period from the 1960s through the 1980s brought about

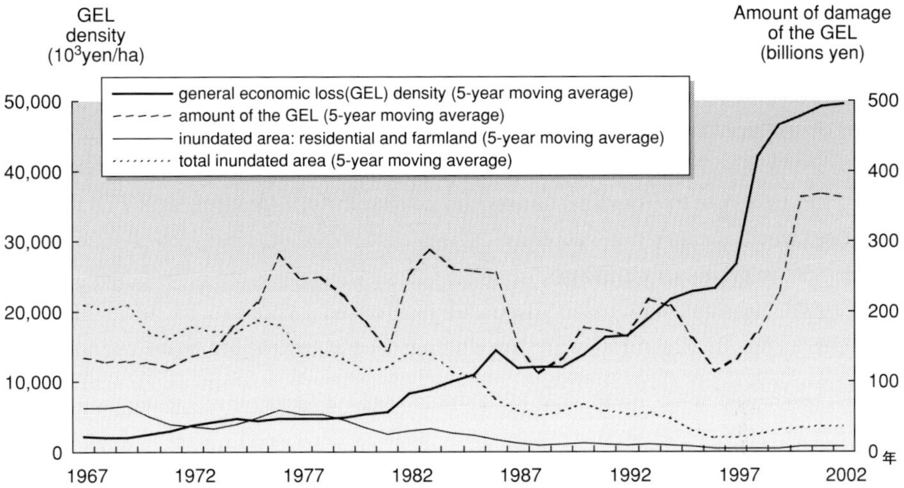

Fig. 4. Trend of General Economic loss density in Japan (MLIT, 2001).

rapid population growth in Japan's metropolitan areas. Without effective laws governing land use, populations and assets became concentrated in flood-prone, low-lying areas that had previously been vacant or used as paddy fields. Currently, 49% of the Japan's total population and 75% of the country's total assets are concentrated in alluvial lowlands, which account for 10% of the land in Japan.

Fig. 4 shows the trend in the totally inundated areas in residential districts, the economic loss of general assets, and the density of economic loss over the previous 40 years. Numerous river improvements have effectively reduced the total number of inundated areas. There has been no decrease, however, in the general economic loss. The solid line indicates that the density of general economic loss per hectare of inundated land in residential districts has been rising in recent years, demonstrating the increasing potential for damage discussed in the previous section.

Exposure to flood hazards has changed drastically through the concentration of populations and assets. Though very rare, catastrophic embankment failures do happen, but the resulting damage can be tremendous, especially in urban areas. This type of disaster is called a Low Probability but High Consequences (LPHC) event. In fact, severe flood disasters caused by embankment failures recently have been common in Japan. The 2000 Tokai flood disaster is an example of how damage increases as a consequence of an LPHC event in urban areas. The embankment failure of the Shin River, a medium-size river, resulted in considerable damage to Nagoya's metropolitan areas and caused the worst general economic loss in 40 years. The area that flooded due to the embankment failure accounted for only 16% of the total flooded area, but the value of the damage was 56% of the total. The economic loss to general assets due to inner flooding of the floodplain was 9 million yen/hectare. On the other hand, the loss due to flooding caused by embankment failures was 778 million yen/hectare.

3.2.2 New types of damage

Social changes have also drastically affected exposure to flood hazards, and new types of flood risks are appearing. Some examples from recent flood disaster events are given below.

1) With changes in the social structure of post-industrial society, urban space is becoming denser and more complex. The urban facilities, information systems, and networks that are now being built are particularly vulnerable to flood damage. Recent urban floods are examples of a new type of disaster that causes a new type of damage in urban areas. For example, in the 2000 Tokai flood, subway lines and stations, building basements, and underground machine rooms for huge storm water pools were inundated. ATM machines were also damaged. Around 100,000 cars were damaged and the insured loss was a record 54.5 billion yen for 58,000 cars. In 1999 and 2002, floodwaters from small and medium-size rivers in the Fukuoka metropolitan area damaged underground shopping malls and the basements of buildings around Hakata Station.

2) Japan has a rapidly aging population. Many elderly people have difficulty to respond to emergencies due to disabilities, limited access to information, etc. In the 2004 Niigata flood disaster, many elderly people drowned in Sanjo City.

3) LFCS measures have created complicated networks of highly artificial river channels that adversely impact the natural environment and wildlife habitats. Until the 1970s, the primary concern was reducing flooding and securing adequate water supplies for industry and urban residents. Natural river courses have been converted into man-made channels for discharging floodwaters. With the advent of more diverse social values, including greater global environment awareness, environmental deterioration, such as the degradation of rivers, landscapes, and

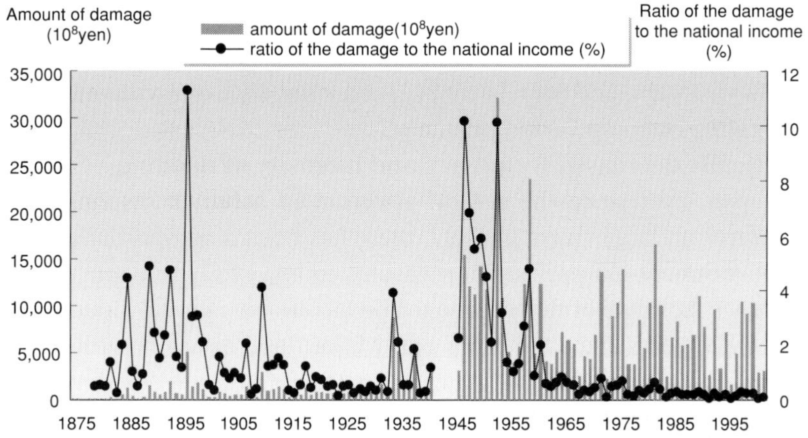

Fig. 5. Trend of flood damage in Japan (MLIT, 2001).

wildlife habitats, is viewed as a type of damage typical in the late 1970s (MILIT, 2004).

4) Since the early 1960s, high economic growth has been accompanied by degradation of the water cycle and soil resulting from the construction of flood control and debris dams and the removal of gravel. Coastal erosion is another problem everywhere in Japan. For example, decreasing silt supplies at the mouth of the Shinanogawa, caused by the construction of a flood diversion channel known as "Ookozu-Bunsui," has led to significant erosion of the coastline west of the main channel. The shoreline eroded 250 m in the 25-year period from 1921 to 1946. Even after restoration work was completed, erosion continued since then (Sunamura, 1996).

3.3 LPHC type flood risk and resilience of society to flood risks

Recent flood disasters show that flood risk in Japan is characterized by LPHC type risk created by natural and artificial environments. In this section, social resilience to flood risks is described, focusing on LPHC type flood risk.

3.3.1 Hazard mnagement

The resilience of society to flood risks refers to our capability to cope with flood risk by means of integrated management of hazards, exposure, and damage. As shown in Table 1, residents, local communities, and governments may reduce the flood risk by implementing measures that prevent certain events from occurring. Such measures include (1) reducing the runoff in catchments, the runoff concentration into river channels, flood-wave propagation downstream, and inundation; (2) reducing exposure vulnerability; and (3) Mitigating or compensating damage.

1) Drastically decreased frequency and intensity of flooding

In Japan, ever since the central government assumed responsibility for flood control on large rivers in 1897, there has been disproportionate emphasis on government-led preventive measures to use LFCS to control flooding. Fig. 5 shows the value of flood damage and its ratio to national income over the period of 120 years since the Meiji era. Major flood disasters with more than 1,000 casualties continued to occur until after the Second World War. Intensive use of LFCS coincided with the growth of the economy. LFCS measures have effectively controlled flooding. Today, most of major alluvial lowlands are no longer vulnerable to repeated flooding and the ratio of flood damage to the national income has fallen to less than 1%. Floods, however, still cause damage valued at 1–1.5 trillion yen per year.

2) Emerging new risks caused by intensive LFCS measures

However, intensive river improvement work based on LFCS has become the norm and one of the main factors behind such new risks as the increased potential for catastrophic flooding, more complex flooding in urban areas, deteriorating river environments, degraded soil and water cycles, and weakened disaster prevention efforts by local communities and residents.

To solve these problems, a new paradigm for flood control was implemented in the 1980s to (i) conserve and create rich natural river environments, scenery, and eco-systems; (ii) promote not only LFCS measures but also various measures in catchments to reduce flooding and damage; (iii) reduce flood risks with the participation of residents; and (iv) support lifestyles associated with rivers (MLIT, 2004).

3) Integrated approach to reducing LPHC-associated flood risks

The basic approach to mitigating LPHC flood risks should be strengthening embankments to prevent failures. However, such work is expensive and time-consuming as the embankments are longer and continuous (MLIT, 2004). To permanently reduce LPHC flood risk, this conventional approach should be shifted in favor of a comprehensive and integrated approach.

Before LFCS were introduced in the latter half of the 19th century, the intensity of floods was reduced through a combination of both physical and social measures. For instance, the force of the floodwaters was reduced by flood forests, while areas prone to flooding implemented retarding basins and second levee systems that redirected flood water to vulnerable low-lying areas.

Table 1. Measures to mitigate in flood risk in occurring process by local resident, local community, and government.

Flood disasters in occurrence process	Purpose	Government	Local Community	Residents
Rainfall ↓ (Catchments) ↓	**Reduce precipitation**	Activities to prevent: the rise in sea level enlargement of typhoon heat island	Activities to prevent worsening global environment	Activities to prevent worsening global environment Improvement in a lifestyle
Runoff ↓	**Reduce runoff generation & runoff concentration**	Disaster-resilient regional development without destructing water environment	Storing rainwater	Knowing the land condition & flood characteristics Water retention on the site
↓	**Reduce flood wave propagation**	Dam Retarding basin	Storing rainwater	Storing rainwater on the site
(River channel) ↓ Inundation	**Inundation**	Flood control works ·bank ·urban drainage system ·pumping station	Ring levee Flood prevention forest Flood fighting (patrol,piling sandbags) Double embankment Floating weir	Trees around the residence Tax burden
↓ (Exposure) ↓ Damage	**Decrease vulnerability of Exposure**	Disaster-resilient social system ·land use management ·taxation system ·flood disaster prevention planning ·hazard map ·disaster education	Disaster-resilient community ·flood-fighting groups ·agree on mutual aid ·network in emergency ·transmit legends and advice ·participate discussion about safe community	Preparedness for flood disaster ·move to safety place ·ensure flood-resilient way of life; banking,flood-resilient house, insurance, and place for household goods & valuables ·prepare materials in case of emergency
↓ (Social systems, etc.) ↓	**Reduce damage**	Providing disaster information Evacuation warning system Assisting volunteer group for disaster preventive activities	Flood fighting activities ·rescue mutual aid ·stop spreading damages ·gather and share information	Reaction to the emergency: ·pay attention to disaster information ·prepare for inundation and rain conditions ·evacuate
Damage	**Reduce damage**	Emergency management Restoration work Financial/Material aid Victim aid system Flood insurance Social/Economic recovery policy	Mutual aid First aid/Recovery Rescue Financial/Material Aid Volunteer/NPO/NGO	Reaction to just behind the disaster

Damage:human damage, economic loss, handicap for live, psychological damage, aggravation of natural environment
measures:deter & reduce damages, hard/soft, long/short-term, permanent/emergency, preventive/reform measures

Overflows were minimized through the use of sandbags filled and placed by the flood brigades of local communities. Small water retention facilities in the house yards delayed the runoff and reduced the peak flood discharge. These were some of the measures available to local communities and residents.

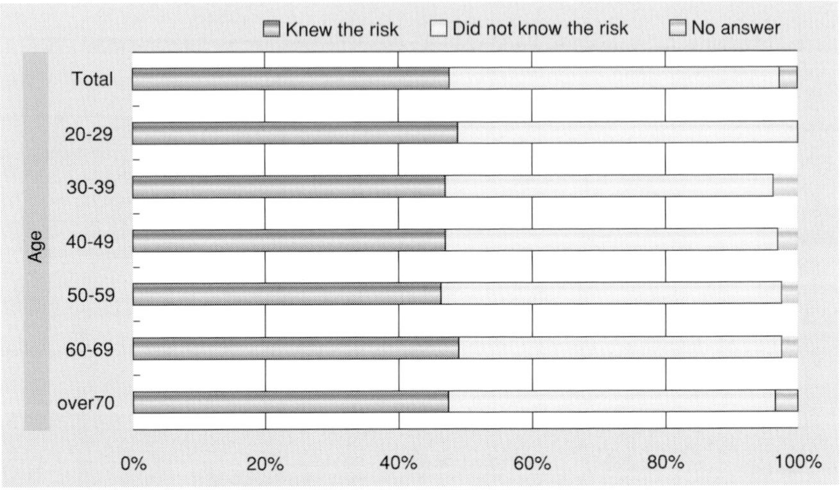

Fig. 6. Flood Risk Perception in the 2000 Tokai flood stricken area (T. Sato *et al.*, 2001).

3.3.2 Exposure management

1) Reducing vulnerability to flood risk

As mentioned above, the quality and quantity of damage caused by flooding has changed as a result of urbanization and more intensive land use. The potential for damage has increased along with the possibility of catastrophic damage from unexpected scale of flooding. New types of flood damage have also emerged. Few existing measures can reduce the vulnerability of the floodplains, and these measures are not effectively regulated in flood-prone areas. This is why urban areas potentially can suffer catastrophic flood damage.

It is the time to think seriously about managing decreasing the risk of flooding associated with urban planning. It is possible to decrease the potential for damage in flood-prone areas by managing land use through construction regulation, etc. Formerly, land use management reduced the potential for damage by placing houses on natural levees, in higher locations on floodplains, raising the foundations of homes, etc.

2) Worst-case risk scenarios

One of the lessons of the 2004 Niigata flood disaster is that the regional disaster plans and flood prevention plans developed by local governments must take into consideration the possibility of the embankments failing in a major flood. However, those responsibility for flood-disaster management in Niigata never considered such a possibility of embankment failure, so the measures taken were not always adequate. Moreover, many residents were not aware of the risk of LPHC type flooding and most had not implemented appropriate measures (Sato, 2006). Few local governments had emergency plans to deal with catastrophic flooding. Worst-case risk scenarios need to be explicitly incorporated into their disaster prevention plans.

3) Raising the residents' awareness to the vulnerability of their habitats

Fig. 6 shows the perceptions of residents regarding flooding prior to the severe 2000 Tokai Flood disaster. This survey was taken in the area stricken by the 2000 flood. Only 48–50% of residents of all ages knew of the vulnerability of their area to flood disasters (Sato, 2003). Residents need to raise their awareness of disaster prevention measures in order to decrease the vulnerability of their homes.

3.3.3 Damage management

1) Accepting flood risk

Reducing flood damage requires strengthening resilience of low-lying alluvial lands to unexpected scale of flood hazards. People need to understand that low-lying land cannot attaint "zero flood risk", and that they are to be ready for taking an acceptable level of flood risk. It is a myth that people always demand "zero flood risk". A survey taken in a flood prone area showed that 30% of the respondents accept the risk of flooding below the level of the tatami mats (straw floor mats) in their homes in every one to thirty years.

At one time, a culture for coping with local flooding was nurtured in local communities. For example, "Mizuya", which are annexed buildings specifically designed for use in flood emergencies, were built on higher ground and stocked with preserved food and a boat for use in an emergency evacuation center for the community. In those days, people also had the wisdom to mitigate their damage by themselves. For example, they used tatami (straw) mats to raise important belongings above the water level, and sealed the gaps between sliding doors with newspaper to keep out the water.

2) Reducing economic losses

Until recently, damage management placed priority on reducing the loss of lives by promoting the evacuation of residents. There was less interest in reducing economic loss of social infrastructures, and cultural and environmental assets. However, recent flood disasters have revealed the potential for damage from LPHC events in urban areas. It is now time to prepare for catastrophic economic losses caused by LPHC events by managing not only hazards but also exposure.

One interesting aspect of flooding is that the intensity and frequency can be controlled spatially. The old feudal government sometimes varied embankment heights to protect economic and political centers. These old ideas give us new insights into modern problems. For example, broad indirect damage can mitigate catastrophic economic losses by minimizing the flood risk over the entire catchment. This could be done by artificially or politically controlling the hazards spatially. However, spatial management of flood hazards has long been excluded from consideration under a current political system of local autonomy. Such measures may be controversial in certain regions, but they are worth considering given the potential catastrophic hazard. In fact, reports by the Niigata and MLIT Committees have touched on this idea (Niigata, 2005; MLIT, 2005).

3.3.4 Promoting disaster prevention activities by local communities

Flood brigades organized by local communities play a major role when emergency or disaster prevention activities require large numbers of people. Such activities include sandbagging and recovery work. It should be noted that these local flood brigades play a particularly important role immediately following the initial stage of flooding and until support is received from other regions. For example, during the 2004 Niigata flood disaster, at the initial stages of flood-combat operations, brigade members accounted for 30% of all individuals acting on behalf of the regional disaster prevention organizations.

In the past, communities had a spirit of mutual assistance and preparation regarding flood risk. However, the rapid decrease in the number of floods in recent years has reduced public experience with flood emergencies. Moreover, when the danger from flooding is inevitable, governments ask residents to evacuate via public warning systems. This approach ensures an appropriate level of safety. The result, however, is a weakening of the disaster-prevention activities of local communities. Urbanization has also led to the deterioration of local community relationships and increased the number of new residents who are unfamiliar with flooding vulnerabilities. Local residents and communities have left their safety in the hands of the government. According to our survey, even in Niigata, one of the most active regions in Japan for community flood-prevention activities, 50% of the residents were not aware of the activities of the local flood brigade (Sato *et al.*, 2006). Disaster prevention activities by local communities should be promoted in a new social scheme that includes coordinating new forms of social networks, utilizing modern communications technologies, etc.

4 Concluding Remarks

The characteristics of flood risk in Japan's urban areas, including components and structure, have changed drastically, both quantitatively and qualitatively, resulting in LPHC-type flood risk. Those are:

Changes in the characteristics of components

1) Flood hazard: Extensive development of LFCS has resulted in notably fewer and less intense floods. LFCS have changed the characteristics of flood runoff and made possible catastrophic flooding as a result of embankment failures. Cascading flood hazards are notable in urban areas.

2) Exposure: Growing populations and accumulations of assets in flood prone areas and more extensive land-use have increased the potential for damage. More extensive land-use and changes in urban and social structures have increased exposure and vulnerability and created new forms of flood risk.

3) Damage: Increased exposure in urban areas has made catastrophic flood damage more likely and changed the nature of the damage. Highly developed LFCS have deteriorated the river environments, soil, and water circulation system and weakened the disaster prevention activities of local residents and communities.

4) Social resilience to flood risk: The powerful LFCS type hazard control measures implemented by governments have become mainstream flood control policy. Extensive implementation of these LFCS type measures has produced the side effects mentioned above. "Zero risk Myth" has spread among residents and they have left their safety in the hands of the government. This has been accelerated by the decline of local communities under urbanization. We should lay the groundwork for re-implementing flood preventive activities by local communities and residents. Increased concern about river environments has led to river improvements that conserve the natural environment

5) New LPHC type flood risks: The interaction of the changes in flood risk components has increased LPHC type flood risk in urban areas. This is the most important issue to be solved. However, few measures are currently available to reduce this type of flood risk. Not only must the embankments be strengthened to prevent failure, the powerful but limited LFCS approach must be augmented with a comprehensive and integrated flood risk management system that takes into consideration the process and structure of flooding while promoting various measures to reduce flooding, vulnerability, and damage. Such a system should incorporate all stakeholders, including governments, local communities, and local residents.

Long experience with flood disasters has given Japan an advantage in understanding the positive and negative consequences of disaster risk. To achieve a sustainable reduction in LPHC flood risk and avoid undesired risk caused by flood risk reduction measures, we need integrating different kinds of alternatives. We should also respect the diversification of values among people, make an adequate choice of effective and efficient methods for allocating resources, and encourage disaster prevention activities by residents, local communities, and governmental authorities.

References

Japan Society of Civil Engineers (JSCE), (2004): Early Report on the 2004 Hokuriku Heavy Rainfall Disasters, http://jsce.nagaokaut.ac.jp/ (in Japanese)

Japan Metrological Association (JMA), (2005): Report on the Climatic aberration 2005, pp.383 (in Japanese)

Ministry of Land, Infrastructure, and Transport (MLIT), (2000): 2000 Tokai Flood Disasters (in Japanese)

MLIT (2001): Flood Statistics, 2000, pp. 515 (in Japanese)

MLIT (2003): 2003 White Paper of MLIT (in Japanese)

MLIT (2004): http://www.milt.go.jp/ (in Japanese)

Niigata Prefecture (2005): Report on the 2004 Niigata Flood Disasters (in Japanese), http://www.pref.niigata.jp/

Sakaguchi, Y. *et al.* (1976): Geomorphology in Japan—its Formation and Characteristics, *Kagaku*, **46**, 223–234, Iwanami-Shoten (in Japanese)

Sato, T. *et al.* (2003): Residents' disaster perception and their preparedness: 2000 Tokai Flood Disaster Case, Proceedings of Second Asian Symposium on Risk Assessment and Management, 40–50 (in Japanese)

Sato, T. (2002): Features of Urban Flood Damage in the 2000 Tokai Torrential Rains, *Major Disaster Survey*, **38**, NIED, 99–161 (in Japanese)

Sato, T. (2005): Better Integrated Flood Risk Management-Features of Flooding in Japan, *Hiyoshi-Bulletin of Keio Univ.*, **15**, 25–38 (in Japanese)

Sato, T. (2005): Flood Disaster Risk in Mega-Cities and Its Management, *Japanese J. of Risk Analysis*, 15-2, 43–49 (in Japanese)

Sato, T. (2006): Characteristics of Flood risk in the 2004 Niigata Flood Disasters, *Major Disaster Survey*, **40**, NIED, 33–78 (in Japanese)

Seo, K. and T. Sato (2003): *Environmentally friendly flood risk management*, Environmental Economics and Policy Studies, **8**, 94–104 (in Japanese)

Sunamura, T. (1996): Change of Coastal Landform by Civil Engineering Works, 137–156, Changes of the Japanese Coast, Kokon-Shoin (in Japanese)

Takahashi, Y. *et al.* (1960): *Arguments about Disasters*, Keiso Pub. Co., pp.349 (in Japanese)

Takahashi, Y (1971): *Transformation of National Land and Flood Disasters*, Iwanami Shoten, pp.216 (in Japanese)

A better integrated management of disaster risks: Toward resilient society to emerging disaster risks in mega-cities, Eds., S. Ikeda, T. Fukuzono, and T. Sato, pp. 41–56.

Integration Framework of Flood Risk Management: What should be Integrated?

Kami Seo

1 Introduction

The concept of integrated risk management was not widely known until very recently in Japan. Both government and lay public take it that only the central government should be responsible for the risk management of natural hazard. The Japanese government has mainly relied on structural measures of risk reduction, which directly control the effects of natural events. Structural measures have taken because, first, those measures brings drastic risk reduction effects, at least for a while; second, they do not need communication and cooperation with other units including local governments, communities and lay public.

During the last decade, however, some situation changes have forced us to reconsider the system. One of those changes is financial difficulty in Japan. Long deflation and expected tax income reduction made repaying the national debt difficult, and the large expenditure for the public construction has been criticized from the cost-benefit (B/C) point of view. The second is the public awareness of environmental problems. Rivers are not only 'evils' that bring disaster; they also provide freshwater, fish, rich soils and all the amenities of the waterfront for humans, and habitat for all kinds of aqua life and birds. Large constructions disturb the ecosystem in the rivers, and they often decrease the amenity value of the river. Those non-marketable values are also a part of the construction costs of the structural measures, i.e. public awareness of the environmental values increases the social costs of the construction of a dam. The third is physical difficulty. City areas in Japan are densely utilized; therefore, getting extra land for construction is difficult, if it is not impossible. The fourth is public apathy to the risk. In Japan, we have often very heavy rains (as much as 50 mm in an hour and sometimes 100 mm in an hour). Since most of the cities in Japan are placed on the flood planes of rivers, Japan is a country with a high risk of flood disaster. Therefore, people are used to being aware of the risk from flooding and have personal and regional "experience" to cope with flooding. Unfortunately (or fortunately), with the frequency of flood decreaseing, these experience are being lost.

Such changes have provided the chance to reconsider the national policy of flood-risk mitigation. Integrated risk management aims for lower risks and lower non-marketable costs within a limited budget. Current risk mitigations are planned separately by kinds of hazard, by mitigation measures, and by organization. Consequently, if the total risk is reduced in the proper costs is not clear. Unclear risk-cost balance makes it difficult to reach national agreement as to what can be defined as a "sufficient level of safety" for our society.

Pursuing cost-effective risk management through optimal resource allocation to different types of measures is one of the key concepts of integrated risk management. Integrated risk management is basically the integration of different types of hazard mitigation measures, including both structural and non-structural, and pre-and post measures.

Change in resource allocation is necessarily accompanied by change in the responsible units of implementation. Large technological measures are usually introduced by the central government. Many other risk-reduction measures, however, are better provided by local governments, communities, and property owners. Moving to integrated risk management, therefore, is necessarily accompanied by a change in the share of the burden from the central government to local governments, and from the public sector to private sector.

Decentralization of those in charge allows flexibility in the integrated risk management system. Large-scale structural measures, provided by the central government, are necessarily distributed following the same criteria throughout the country. A nationwide system is equitable; however, it is not always efficient because both natural and social environments vary from region to region. Different social and cultural environments, as well as natural environments, require different methods of risk mitigation. Some can be more risk-averse than others, and some put a higher value on the natural environment of rivers. Risk management by local governments or communities is expected to be more sensitive to the local circumstances. Also, a decentralized decision system makes public involvement easier, which is expected to improve public ability of managing natural hazards.

2 Structural Measures in Japan—What is the Problem?

Japanese flood-risk management after the Meiji period has heavily relied on structural measures such as continuous banks and dams, which have successfully lowered the frequency of flood disasters. In general, however, relying on only a few measures is necessarily becoming more and more inefficient in terms of risk-cost effectiveness because the marginal cost of a measure for risk reduction increases with the progression of the coverage. Adding a meter of bank on an existing one-meter bank is usually less expensive than adding one on two-meter bank. However, benefit is usually larger in the first raise than in the second one. Therefore, cost efficiency of risk mitigation measures necessarily diminishes (Fig. 1).

Pursuing further risk reduction by structural measures is being more and more difficult both in mega-city areas and in rural areas in Japan. In city areas, there are physical difficulties, as previously mentioned. Another factor that makes cities vulnerable to extreme natural events is low public awareness of risk. People always have difficulty remembering natural disasters if they are not frequent; for example, events that occur once in a hundred years (Huntington and MacDougall, 2002). People in the city areas are especially unprepared for natural hazards, because first, flood in mega-cities are less frequent thanks to the improved technology. Second, people in cities do not know the nature of their homelands because many of them come from the countryside, and their residential time in the city tends to be short. Consequently, losses due to floods often become larger. In the Tokai Flood in 2000, the area affected, including Nagoya, the third largest city in Japan, for example, the economic loss was as large as 730 billion yen (6.6 billion US dollars) (NIED, 2002). There is a dynamic relationship between flood reduction by structural measures and vulnerability of the city area; structural measures against floods lower the frequency of hazardous events, which attracts more people and accelerates development. Then the area requires a higher level of structural measures against floods. This spiral necessarily raises both the risk of floods and the cost of countermeasures (Seo and Sato, 2003).

In city areas, however, the cost-benefit balance of construction is better compared with in rural areas. After the Tokai flood disaster, a typical city flood disaster in 2000, for example, 61 billion yen (approximately 550 million US dollars) was used to improve structures around the Shonai river, but that cost is relatively moderate compared with expected loss in the future. In rural areas, on the other hand, physical restriction is not so tight; however, construction costs of large structures are much higher compared to the benefit, because of the small number of residents. After the earthquake in Niigata, rural area in Japan, in 2004, restoration costs of the infrastructure and public goods are estimated to be more than ten million yen (80 thousand US dollars) per person in small villages on the mountainsides. As hundreds of rural areas in Japan face a gradual population reduction, the situation looks as though it is going to get worse. These B/C inefficient undertakings will not continue to be accepted under the financial status of today's Japan.

Sometimes, it is not clear if the *risk* is reduced by structural measures. In

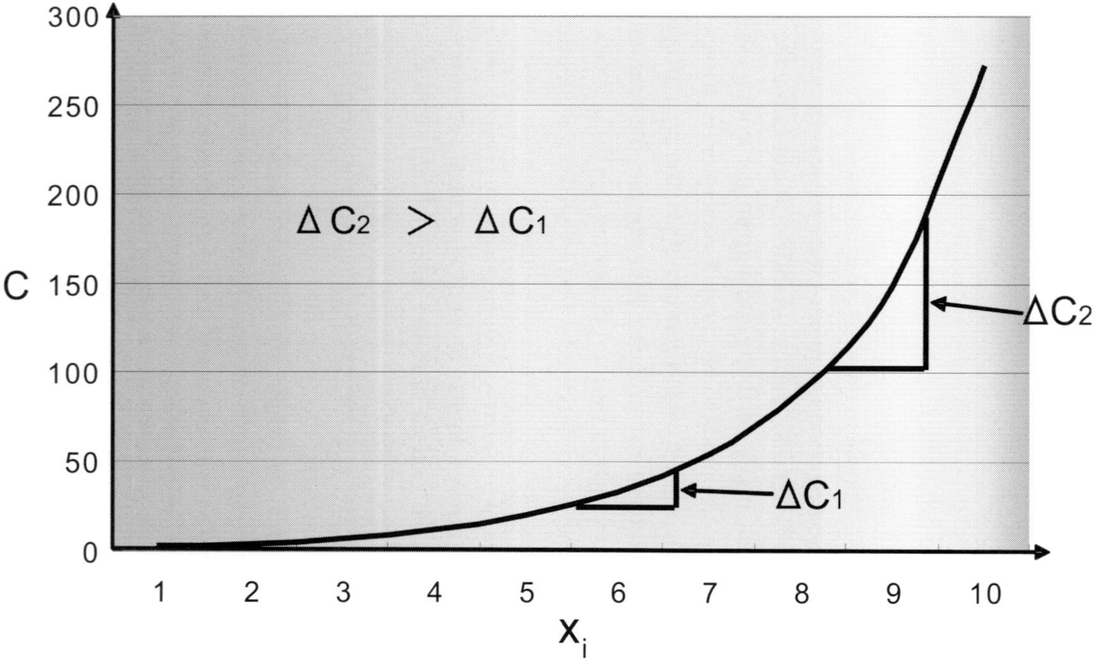

Fig. 1. General pattern of the measure improve (X_i) and its cost (C).
Under the same technology, the marginal cost of a certain measure usually increases with progression of the coverage, i.e., ΔC_2 in the figure is often more expensive than ΔC_1.

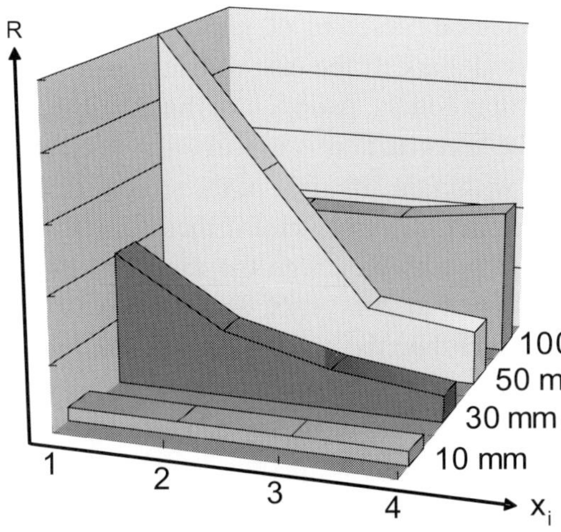

Fig. 2. Risk(R) with the progress of risk reduction measures (Xi).
The vertical axis represents risk of flood and the horizontal one represents the level of a measure: let us imagine the height of the bank. Relationship between the height of the bank and risk reduction effects is not linear as people intuitively expect. In case of frequent but not heavy rain (10mm/houre) even a small bank may not be necessary. In case of moderately heavy rain (30mm/houre) the low bank may be enough. In the rare case of heavy rain (50mm/houre), risk will be reduced by higher bank. Then what will happen in extreme case (100mm/houre)? This case is very rare, which may happen once in a hundred years. Thus the total risk of this case is not as large as the risk of 50mm/houre case, because of its low probability. However, the magnitude of hazard of each event is much larger. In this case, bank is often useless, or even worth; the reason is because, larger bank collapse gives larger damage than smaller bank collapse.

general, the lower *frequency* of floods does not mean a lower *risk* of flood. Continuous banks increase the quantity of river flow because water that used to flow out at the upper stream is retained by the river. In fact, the quantity of water in Tonegawa in Tokyo was 3.8 thousand square meters in 1900 and it became more than three times greater as 14 thousand in 1950 because of the 'Great Wall' like bank construction (Takahashi, 1971). Larger rivers cause more catastrophic flood disasters. Damage from flooding after the completion of the bank can be much greater than before the construction. In other words, a trade-off exists between the frequency and the size of hazards.

Large constructions are often decided upon by the government after a flood (Platt and Rubin, 1999). They are usually planned to help prevent further floods. However, if the decision is right or not, in terms of the cost-benefit balance, are not clear. For example, a town had a bank that resists 50 mm/hour rain, but is attacked by a flood caused by 60 mm/hour rain. Suppose the loss of disaster was 100 million dollars. If the cost of bank improvement to resist 60 mm/hour rain is 30 million dollars, people will think that the construction is a good deal. However, the construction is useful only when they have just 50–60 mm/hour rain again within the life time of the structure. Up to 50 mm/hour rain, the constructio n is not necessary and if there is 100 mm/hour rain, the bank is not only useless but may enlarge the disaster. If the probability of 50–60 mm/hour rain is once in 200 years, and the life of the structure is 50 years, if the benefit is larger than the costs is not obvious. In general, one unit of structural measure reduces certain levels of flood disaster; however, it may not be necessary in most cases and/or it may be useless in the case of a really large natural event (Fig. 2). However, people tend to assume that *the risk* is automatically reduced by the construction.

3 Integration of Different Types of Measures

3.1 Hard and soft, pre and post

Our society has relied on structural measures to prevent flood. The purpose of integrated risk management, however, is not flood prevention but reducing human and/or financial loss through flooding. Preventing floods can be one of the ways, but in itself it is not the main purpose. We may accept flooding once in a while if the loss is low enough.

As is often discussed, risks of natural hazards are at the intersection of extreme natural events and the human system (Burton *et al.*, 1993). Therefore, in order to reduce flood losses, preventing flooding is not the only way. The pathway from the natural event to the endpoint—loss of flood—is not simple, and measures to reduce risks are varied. We analyzed the structure of the pathway of flood hazard in mega-cities in Japan by drawing a flood causal model (Fig. 3), originally constructed by Hoheneser *et al.* in (1982). In Japan, with population growth, high-risk areas in cities become densely populated (at the upper left of Fig. 3). At the same time, suburbs of the cities are developed from agricultural lands to residential areas and that lower the penetration of lands. Since Japanese mega-cities are placed on flood plains of large rivers and their suburbs are at the upper streams of the river basin, a change in the land coverage of suburbs often presents extra loads to rivers. Then unusual natural events are more likely to cause river overflowing. Also, cities are increasingly flooded, without river overflow, by internal water. This is because the capacity of drainage becomes less sufficient since penetration of the lands rapidly becomes smaller than expected, and concentrated land use impedes magnification of drainage capacity. Finally, flooding causes human death and property loss (at the right of Fig. 3). In order to reduce risks, one of the causal chains above needs to be interrupted.

- Prevention

Flood prevention by hard measures is one of the most utilized ways in Japan. Preventing flooding is a direct way to reduce risks. The problems of relying too much on preventative measures have already been discussed.

- Information and Education

Integrated risk management focuses more on non-structural measures at the lower reaches of the causal flow, improving social vulnerability and reducing human and financial losses in the case of flooding. Those measures include emergency alarm systems, hazard maps and web education systems, land use regulations, and insurance.

An emergency alarm system is the last system to save lives in the case of flooding. It is simple and has a relatively long history. However, the timing of the alarm and evacuation is not always easy. Obviously it should not be too late, but if it is too early, it will not be taken seriously. Also, too early evacuation always attracts looters to the region.

In order to reduce human losses in the case of flooding, education about hazard is important. In cities, many people have no experience of floods, and they do not know the fear of flooding. Sometimes people die by remaining in basements during flooding. Another group of people who are use to flooding, often take quick action: moving goods from the first floor to the second and going to higher places with some water, snacks and money. Thus, risk education and emergency alarm systems are expected to effectively reduce human losses even if those systems do not reduce flooding themselves.

- Land use regulation

Change in land use is also one of the traditional non-structural measures that drastically reduce human and property losses. As is often mentioned that "water remembers the way". Water reaches lower places by always taking the same paths. Therefore, if society pays more attention to the land use of those places, flooding will not be followed by losses.

- Post-event planning

Post-event planning is an important topic of integrated risk management. Today, property losses through flooding is expanding and this can be covered by insurance. In many cases in Japan, constructing structural measures is much more expensive than the expected risk reduction. In other words, prevention needs more social resources than restoration after the event. In those cases, risk transfer through insurance is more appropriate than pre-event risk reduction.

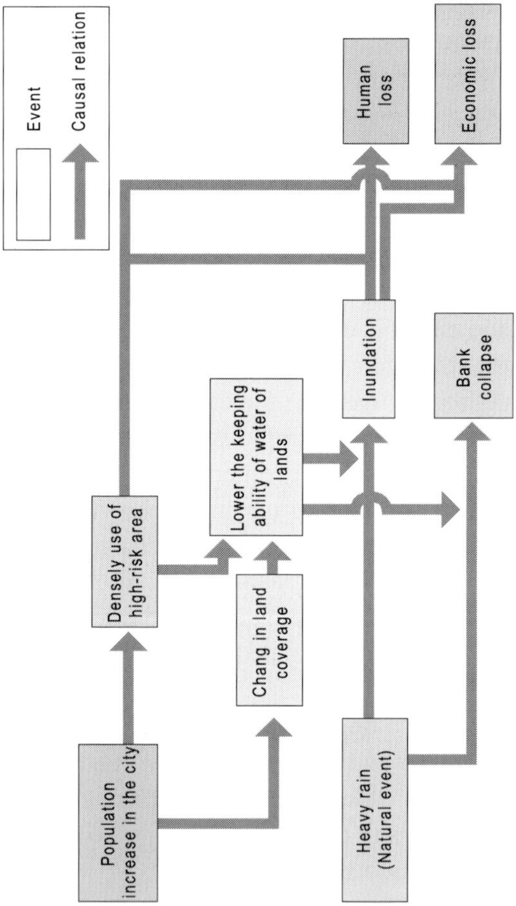

Fig. 3. Causal relationship between natural/social events and loss of the events.

Risk mitigation measures are varied, including structural and non-structural, technological and social, prevention and restoration. Integrated risk management is the optimal allocation of human and monetary resources among different kinds of measures. If measure A is more cost effective than measure B, re-allocating more resources from B to A is efficient, which means reducing more risks within the budget.

3.2 Integrated management of multi-hazards

Discussed above is the efficiency of the integration of different types of mitigation measures against a single hazardous event. Theoretically, the same discussion is possible among mitigation measures of different kinds of hazardous events, such as floods, earthquakes, and criminality. Currently, different types of hazards are managed differently, mainly for historical and institutional reasons. As both target level of risks and the budget of risk management is set by each hazard, marginal risk reduction cost can be higher in one kind of hazard than another. In those cases, monetary resource re-allocation across different kinds of hazards is efficient.

Also, integration of different types of hazard management is efficient because it can often curtail redundancy. For example, the infrastructure of an information system can be used for plural hazards—or the system should be constructed to be adaptable to multi-hazards. We found one unsuccessful case in the Tokai flood—some shelters and emergency storage were submerged (Seo and Sato, 2002). Those shelters are provision against disasters, but it is easily imagined that an earthquake was mainly assumed, because those shelters were constructed soon after the catastrophe of the Hanshin earthquake. Shelters were constructed in low places. If they had been constructed by the budget of flood-risk prevention, they would have been on high ground, but could have been more vulnerable against earthquake. The idea of multi-hazard will be helpful to construct a facility that is tolerable against both hazards.

As will be discussed, communities are expected to play an important role in integrated risk management. The idea of multi-hazard management will be important especially in community level risk management, because human resources are limited in communities. Recently in Japan, with crime on the increase, the role of the community in the prevention of crime is expected. Systematization of the community, not only against crime but also against multi-hazard, makes it less vulnerable.

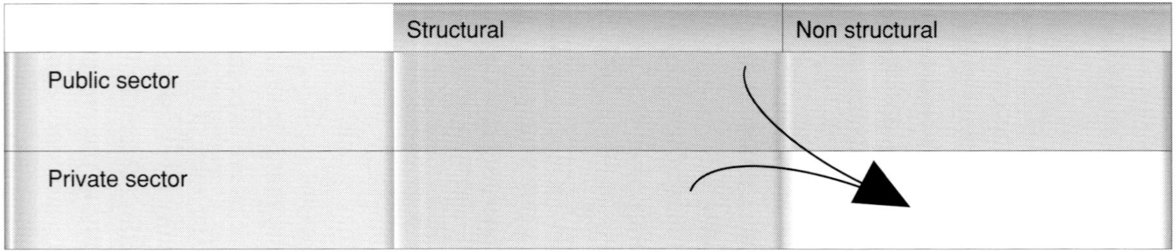

Fig. 4. The type of measure and the responsible units of implementation.
Change form the hard to the soft in the risk reduction measures often means the change from the public sector to the private sector in the management units.

4 Integration of Different Levels of Risk Management

4.1 The role of Government

Resource re-allocation with integrated risk management necessarily requires changes in the unit of risk management from central government to the local governments, communities, and individuals and from the public sector to the private sector (Fig. 4). The expected role of the government in integrated risk management is different from that in flood prevention by the structural measures. In the latter, the role of government is basically assessing the risk of flooding, and constructing banks and/or dams. The process requires careful observation of natural events, a high level of technology, and a large enough budget but not cooperation with other units. In integrated risk management, the role of government is more varied and complex. The government has to plan how to participate with the local governments and communities. It is also required to plan how to share the responsibility with the private sector. If the government's share or responsibility is too large, the incentive for private sectors to improve their own safety will be discouraged. If the share is too small, however, the poor and the weak may be left in danger.

4.1.1 Information and communication

The importance of the role of government in risk information and education cannot be over-emphasized. The government should appropriately inform people so that they can take rational action. The history of open-information policy is short in Japan. Until recently, many landowners were negative about opening risk information such as distributing hazard maps because risk information may lower the value of their properties. Also, both the government and public shared the same idea that the government could guarantee zero risk; open information was inconsistent with the illusion.

Today, the policy of open-information is widely accepted. Japanese people can get risk-related information through various media if they want it. For instance, we developed a Web system, the Participatory Flood Risk Communication Support System, for education, information and helping risk communication as introduced later in this book.

The argument whether information should be open goes on. However, we have little experience about *how to* open it. Lay people may not pay enough attention to the risk information in daily life; they may feel it is ridiculous to prepare for a hazardous event which may come within two hundred years. If, however, the way of opening information is too sensational, people may overestimate the risk and take unnecessary action, which often causes different types of risks. Studying effective ways of release information and education—implementation science—may be one of the new roles of the government under the integrated risk management.

4.1.2 Informed choice

To reduce risks, the government must make the difficult decision between regulation and informed choice. In general, regulation is easier and a more risk-averse way. Land use regulations against flood, including retarding basins and building regulations, are direct ways for a society to circumvent flood-risk exposure. Keeping a retarding basin in a city area is sometimes costly, but often less expensive than the cost of constructing structural measures.

Regulation, however, is rigid compared with informed choice, and thus is not always supported by the public. Informed choice is often better in terms of cost-benefit efficiency if the public appropriately perceives the risk. It is especially efficient when the value system of the risk-takers and that of policy-makers is different. When the public and policy-makers are sharing the same value system, the public will take appropriate action, even without regulation, only if they are informed. In this case, regulation is not necessary. When the public and policy-makers have different value systems, regulations always lower the risk-takers' satisfaction: if the regulation does not exist, people can take different actions, which increases their benefit, according to their own cost-benefit evaluation. "Satisfaction" of risk-taking people is naturally subjective which can be evaluated only by the risk-taking people themselves.

Effective risk management through informed choice, however, is not always realistic, because many people do not fully understand their own risk of natural hazards. As mentioned, risk perception is not easy for many people, although their lives are inundated with risk information. Informed choice requires education and communication system at the same time, because it is effective only when a certain degree of rationality of the public can be expected.

4.1.3 Insurance

A substantial role by the government is often expected in the restoration after a disaster. Some people in Japan even insist that the government should restore private property as well that was destroyed by the disaster. In fact, after the earthquake in Tottory in the 2000, the local government restored privately owned houses through taxes. The policy benefited the sufferers, but private loss is better being restored by the owners for the same reasons as informed choice is superior to regulation as discussed above. Inflow of governmental money discourages personal efforts of preparation against hazards, and may cause "moral hazards", which increases the total risk of the region as a result. Loss from natural hazards is often large and certainly is difficult to restore privately. However, although the loss is large, the *risk* is not because the frequency is not large. Thus, risk transfer through an insurance system is preferable.

Most of the insurances are usually provided by the private sector in Japan. However, the government is expected to play a roll in insurance related to natural hazards, because the loss from natural hazards is often huge, which can bankrupt companies. If there is risk of bankruptcy of an insurance company, people will hesitate to buy insurance. Therefore, governmental assurance is worth considering for this system. There is another merit in the cooperation of the public sector and the private sector in improving insurance systems. The government can participate in the designing process of the insurance. Carefully designed insurance can encourage public awareness and preparation for hazards. For example, if the premium is risk-based, insurance provide incentives to insured people to lower their own risks because personal efforts will be rewarded with lower premiums. Living in a safer area and other private risk mitigation can be justified by insurance systems.

In the United States, the federal government runs an insurance system, National Flood Insurance Program (NFIP), which is planned to give incentive to private efforts for safety. The Community Rating System (CRS) of NFIP encourages people and communities to protect their own places from floods through a risk-based premium system (FEMA, 2005). Under CRS, if a community improves the regional safety, the insurance premium is lowered. Therefore, communities and people have incentives to lower the risks. In fact, however, a risk-based insurance system is technically not easy because a fair assessment of risks from natural hazards throughout a country requires a significant amount of both human and monetary resources. Also, risk may be too different among people to reflect correctly on the premium (Adams, 2000). This sort of system can be regarded as for the public good, which is not necessarily appropriate to be provided only by the private sector.

As discussed above, integrated risk management sometimes requires efficient cooperation between the public sector and the private sector.

4.1.4 Post event restoration

Planning of risk-related resource allocation often focuses more on the prevention of disaster than on restoration. People find restoration planning before the event to be relatively hard to accept because it means that the risk is not zero. Actually, however, risks of natural hazards cannot be zero, so restoration planning before events is quite important.

Post-event planning is not only a fail-safe system for hazards that cannot be physically avoided by prevention measures. Sometimes restoration after the event is much less expensive than risk prevention. In other words, saving resources for restoration after the event is often more efficient than allocating resources for prevention. Public agreement may be more difficult to get for the policy, not implementing possible prevention measures and saving resources. However, an appropriate balance between measures taken before and after events ultimately lowers risks within the limited budget.

Prevention and restoration by the government sometimes have different meanings for property owners in the target region. Pre-event measures, which lower risks, directly raise the value of privately owned property. On the other hand, restoration is usually designed not to restore privately owned property through government expenditure, because otherwise, taxpayers' agreement is hard to get. The government typically restores only public property, such as roads and public schools that benefit the private sector only indirectly. Therefore, a change in resource allocation from prevention to restoration partly means a change in the allocation of burdens from public to private, and therefore, property owners have incentives to impede this change and tend to overemphasize the importance of pre-event measures.

The restoration of private property through public expenditure seems unfair to taxpayers, as mentioned above. Private property should be restored by the owner rather than by taxes because only the owner benefits from the restoration. However, the restoration of privately owned property is not necessarily unfavorable for taxpayers, if the marginal cost of curtailing risk through restoration, including the restoration of private property, is still less expensive than prevention, which is charged to the public account. Again, the problem of the government restoring private property is the moral hazard. Knowledge that the government will pay for restoration after a disaster discourages owners' personal efforts to lower the risks. Also, if the government compensates a private loss, no one will buy insurance, which disturbs the risk-reduction function of a risk-based insurance system.

The integration of pre- and post-event risk management is the difficult part of integrated risk management. As governmental policy affects public choice, as discussed above, a simple cost-benefit-based design is not effective. The expected role of the government is preventing moral hazards and giving incentives for self-defense. At the same time, the government should help people who cannot surmount risks through their own efforts.

4.2 Role of residents and community

Public involvement is one of the most important but difficult topics in integrated risk management. When moving from governmental management to integrated risk management, public agreement and cooperation is necessary. The reason is because systems such as informed choice and insurance are effective only when a certain degree of rationality of the public can be expected, as discussed.

However, individual ability to process information varied substantially and their value system is quite different depending on the experience, region they live, age, and income. Some people have experienced flooding and know well about that hazard, but others do not. Some obtain information through the Internet freely and others do not. Therefore, reaching an agreement, even among lay people, is difficult. Those differences among people are one of the sources of uncertainty, because the probability of loss in the case of an emargency depends on the action of individual people.

Agreement between policy-makers and lay people is more difficult because many people do not understand about budget constraints. Another difficulty is related to public agreement to share responsibility. Unlike structural measures or regulations, people have to be responsible for their own choices. Public involvement can lower risks, but sometimes people will find the costs imposed on them. In other words, public involvement is necessarily accompanied by costs, charged directly to the public. Therefore, the idea is not always readily accepted.

Traditionally, however, the risk of a natural hazard was managed by each community to a certain level. Individuals used to take expected action as members of the community. The weak in the community, such as the elderly and handicapped were taken care of by other members. The case study of the Tokai flood show that the collapse of the community makes a region significantly more vulnerable (Seo and Sato, 2002). The community should be appropriately positioned in integrated management.

Expected effects of risk communication are the reconstruction of the community and construction of a good relationship between the government and the community. Rapidly spread Internet provides convenient ways for risk communication. Our project developed a prototype of an e-community platform on the Internet for risk governance by a community, and collecting empirical data on risk communication through the platform using the experiment in Shimada City, as discussed in Chapter 5 (Nagasaka).

5 Direction—Concluding Remarks

Today, governments are expected to provide high levels of safety and social welfare under budget constraints. Thus, a simple strategy of relying on only a few measures provided by the central government is not satisfactory because both environments and human value systems (cultures) vary from region to region. Integrated risk management is, literally, rational integration of different types of measures: technological and social, structural and non-structural, and preventive and restorative.

Different types of measures require different levels of risk managers. Integrated risk management requires the involvement of local communities and cooperation between the public and private sectors. It is necessarily accompanied by a change in the role of the government. In addition to providing technological measures, the government is expected to support regional decision-making. Effective information systems, community involvement and alternative ways of allocating resources will be more important in risk management in Japan.

Public involvement is one of the key concepts of integrated risk management. To achieve the optimum balance of risk and cost, involved people are required to share certain expenditures; otherwise, the demand on public goods becomes too large. A system should be designed to bring out the ability of the public to help itself. At the same time, it must provide a high level of safety to those who are not capable of helping themselves.

References

Adams, J., "A Richter Scale for Risk?" (2000): in Julian Morris eds., *Rethinking Risk and the Precautionary Principle*, Butterworth Heinemann, Oxford, UK.

Burton, I., R. W. Kates, and F. G. White (1993): "Hazard, Response and Choice", in *The environment as hazard*, Guilford Press, NY, NY.

FEMA (2005): http://www.fema.gov/nfip/crs.shtm (accessed in Aug. 2005).

Hohenemser, C., R. E. Kasperson, and R. W. Kates (1982): "Causal Structure: A framework for policy formulation", in C. Hohenemser and J. X. Kasperson eds., *Risk in the Technological Society*, Boulder, CO, Westvies Press.

Huntington, S. and K. MacDougall (2002): "Flood Risk", in George Fleming eds., *Flood Risk Management*, Thomas Telford Ltd., London, UK.

NIED (2002): "Report on Urban Flood Disasters and Landslides on September 2000 in the Tokai Regions, Japan", in Natural Disaster Research Report No. 38.

Platt, H. R. and C. B. Rubin (1999) "Stemming the Losses: the Quest for Hazard Mitigation", in H. R. Platt eds., *Disasters and Democracy: The Politics of Extreme Natural Events*, Island Press, Washington, D.C., 69–107.

Seo, K. and T. Sato (2002): "Tokai Disaster as an Urban Flood hazard: a survey report", in "Report on Urban Flood Disasters and Landslides on September 2000 in the Tokai Regions, Japan", in Natural Disaster Research Report No. 38, 163–176.

Seo, K. and T. Sato (2003): Environmentally friendly flood risk management, *Environmental Economics and Policy Studies (Japanese)*, **8**, 94–204.

Takahashi, Y. (1971): Changing country and food Iwanami, Tokyo, Japan.

A better integrated management of disaster risks: Toward resilient society to emerging disaster risks in mega-cities, Eds., S. Ikeda, T. Fukuzono, and T. Sato, pp. 57–87.

Public Preference and Willingness to Pay for Flood Risk Reduction

Guofang Zhai

1 Introduction

Floods are one of the oldest, most frequent, and most severe natural disasters to which human beings are exposed. In Japan, floodplains hold about 49% of the population and 75% of property, and flooding is a serious natural hazard. The need to reduce flood risk by applying not only "hard" measures, such as dams and levees, but also "soft" ones, such as early-warning systems and measures to heighten awareness of flood risk, is thus widely accepted. Unfortunately, the severe economic recession in Japan and public concern with the need to preserve the environment have recently put a brake on investment in large-scale public works, including flood prevention schemes such as dikes and dams.

In these early years of the 21st century, floods have occurred frequently in Japan, causing disastrous loss of life and property. The Tokai Flood in 2000 caused damage amounting to 978.3 billion yen, with 10 dead and 115 injured. The Niigata-Fukushima Flood on July 13, 2004, left 16 people dead or missing, 22 buildings destroyed, 156 severely damaged, 85 partially damaged, 4,022 inundated above ground, and 22,620 inundated below ground. The frequency of flooding provides adequate evidence to support advocates of large-scale flood disaster prevention schemes.

Zhai *et al.* (2003) conducted an efficiency analysis of Japanese flood prevention investment vs. total flood losses including death, injuries, and intangible effects and showed that, since the 1980s, investment has changed from an efficient mode to an inefficient one in terms of both the economic standpoint and total savings on flood losses, including loss of life, with the ratio of total benefits to total costs, elasticity, and the marginal rate of substitution for flood loss decreasing in relation to investment in flood prevention.

The objective of this paper is to answer important five questions by conducting a survey in the Toki-Shonai River region of Central Japan: What are the main public preferences regarding river management in Japan? What is the public flood risk acceptability? What is the willingness to pay (WTP) for flood risk reduction? What are the relationships among the attributes of flood risk reduction like economic damage reduction, human loss reduction, and environmental protection?

2 Theoretical Framework for Estimating WTP

Water quality, river landscaping, flood control, and similar issues have no market price tag. These are called non-market goods in economics. Information obtained by assessing the public's interest in, and willingness to pay for flood control and ecological restoration is critical for municipal planners and policy-makers who desire to vote on legislation according to their constituents' wishes. The numerous techniques available for estimating WTP can be broadly divided into two categories: revealed preference and stated preference methods. The former, such as the travel-cost and hedonic price methods, determine the demand for goods or services by examining the purchase of related goods in the private market place, while the latter, such as the contingent valuation method and choice experiment techniques, measure demand by examining the individual's stated preference for goods or services relative to other goods and services. The choice of method for a study depends on several criteria, including the purpose of the study, availability of data, and particular economic values required (use and/or non-use values). For example, the revealed preference method is still generally preferred for certain types of non-market goods like recreational fishing, while the stated preference method is applied to other types of non-market goods such as estimating discount rates in developing countries (Bateman *et al.*, 2002). Shabman and Stephenson (1996) showed that the contingent valuation method (CVM) produces the smallest mean estimates, implying that CVM is the most conservative estimation technique and is thus least likely to overestimate actual benefits based on the property damage avoided, hedonic price, and contingent valuation techniques for the same study area (Roanoke, Virginia, USA). As highlighted by Daun and Clark (2000), however, CVM has rarely been specifically applied to the estimation of flood control benefits (Thunberg, 1988; Shabman and Stephenson, 1996; Clark *et al.*, 2002). On the other hand, the choice experiment approach has not been applied in the flood risk analysis as far as we know.

CVM involves posing a hypothetical market to a sample of respondents and asking their opinion on the values of public environmental goods or services (e.g., WTP for a change in the supply of an environmental resource) under specified contingencies (Carson and Mitchell, 1989; Freeman, 1993; Bateman and Willis, 2001; Bateman *et al.*, 2002; Zhai *et al.*, 2006b). CVM is the most flexible of the methods available for measuring both direct and indirect monetary benefits of non-market commodities like environmental resources. If respondents answer a CVM question, as assumed by economic theory, the elicited value corresponds to the economic value of the goods (resource) as measured by the Hicksian compensating surplus (Carson and Mitchell, 1989). The estimated value for a natural resource can then be used as an input to a cost-benefit analysis.

The choice experiment approach is based on the idea that any goods can be described in terms of its attributes, or characteristics, and their levels (Bateman *et al.*, 2002). For example, a bus service can be described in terms of its cost, timing, and comfort. Likewise, flood prevention measures can be described in terms of hard measures such as internal and external flooding measures, soft measures such as early warning systems, and concern about the environmental protection of rivers.

WTP for flood and environmental risk reduction may depend on factors such as risk perception, resource limitations, personality (individual characteristics), current risk levels, and acceptability of risks. Whether an individual acts or not depends on whether his utility reaches a maximum, which is strictly confined to the addressed factors. Figure 1 shows the theoretical framework used in this study for estimating WTP for flood risk reduction. The hypothesis for the theoretical framework was tested using CVM and the choice experiment approach.

3 Survey Design and Implementation

3.1 Survey area

The survey area was the Shonai-Toki River basin in Central Japan (Fig. 2). The upper and the lower reaches are called the Toki River and Shonai River, respectively. The main stream is 96 km long. The basin has an area of 1,010 km^2 and is home to about 4 million residents. The upper reaches area is quite different from the lower one in terms of both natural environment (climate, geographical features) and socio-economic patterns (urbanization, population, and property accumulation). Therefore, the Kita ward of Nagoya City, Aichi Prefecture, in the lower reaches area, and Toki City, Gifu Prefecture, in the upper reaches area, were selected as survey areas to examine whether public preferences regarding flood control would differ with location.

Disastrous floods frequently occur in the survey area, e.g., during the last 50 years, Typhoon Isewan in 1960, Typhoon No. 17 in 1971, floods in 1989 and 1994, and the Tokai Flood in 2000. In particular, the Tokai Flood, resulting from heavy rainfall of up to 97 mm per hour with a total precipitation of 567 mm, inundated the Tokai area, including the city of Nagoya, home to 2.1

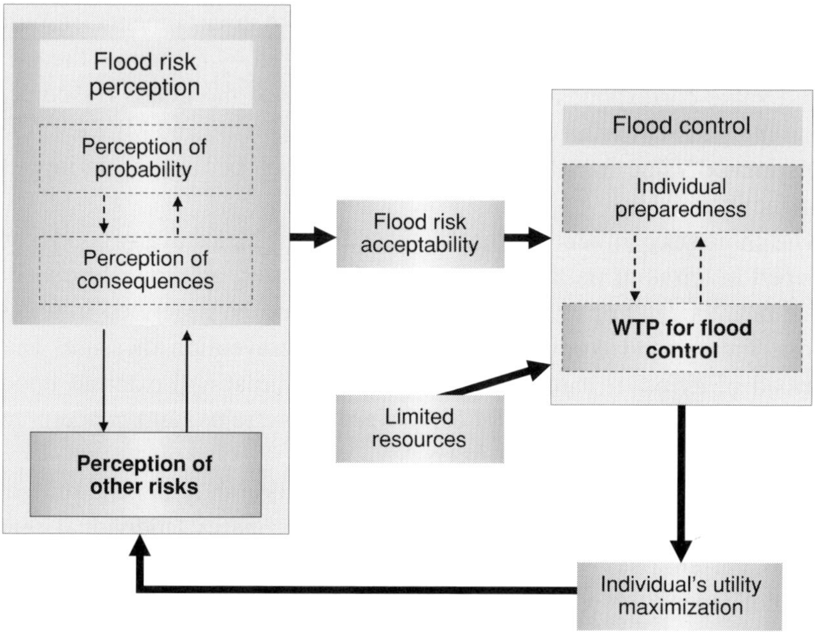

Fig. 1. Theoretical framework for analyzing WTP for flood control.

million residents. This storm, regarded as a "once in over 200 years" event, caused 10 deaths, serious injury to 20 people, and ¥978.3 billion in direct economic losses. It was one of the most serious flooding disasters in Japanese history according to Ministry of Land, Infrastructure, and Transport (MLIT) statistics.

3.2 Survey implementation

For CVM to yield useful information, careful survey design is critical. The survey purpose determines the accuracy of the results and survey mode. Desvousgaes *et al.* (1998) pointed out that there are different levels of WTP accuracy required depending on the end use. These were ranked, in descending order of accuracy, as compensable damage/externality cost, policy decisions, screening or scoping, and fact finding. The ultimate implications of this survey include possibly helping to improve policies for flood control projects in Japan and gaining a better understanding of theoretical issues such as the effect of providing information in questionnaires on WTP. The purposes of this survey thus belong in the two latter-ranked categories. Therefore, a mail survey was considered appropriate, rather than more expensive face-to-face

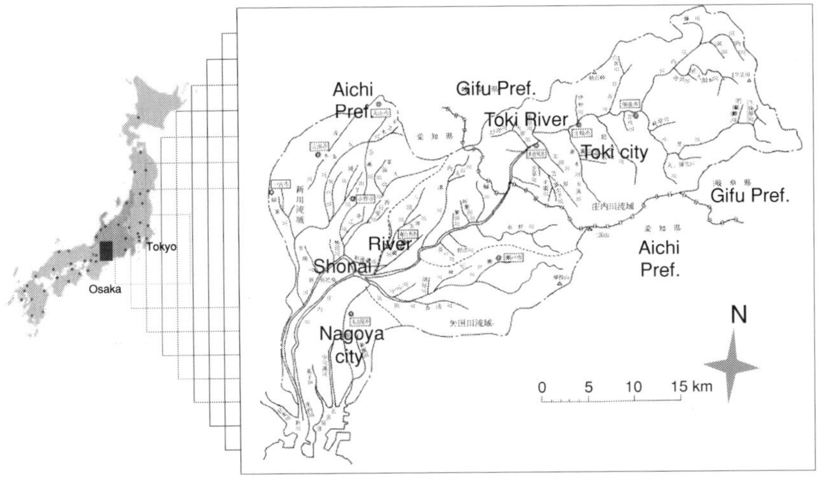

Fig. 2. Location of survey area.

interviews, as proposed by Arrow *et al.* (1993).

The survey was conducted from the end of March to the beginning of April, 2004, before the start of the rainy season (June to September). The survey followed the Total Survey Design Method (TSD), which attempts to achieve an optimum balance across all effort areas. TSD was developed by Mangione (1995) and has been successful in securing high response rates from general and special samples. The survey procedure was as follows.

First, brief descriptions of flooding, the current status of flood control measures, and methods of preventing flooding were presented. In particular, for ease of understanding, illustrations were used to present the relationship between property damage and flood inundation, and to describe external flood control measures, internal flood control measures, and early-warning systems. To provide respondents with a further incentive to answer the questionnaire beyond involving them in cooperative thinking on measures for flood risk reduction, they were asked to write their name and address on the questionnaire if they wanted a summary of the survey report.

Second, in addition to thorough discussion of the questionnaire draft within the project team, six people were asked to formally pre-test the questionnaires. Based on the information returned, the questionnaire was revised.

Third, the questionnaires were sent to 1,000 selected households by mail with a covering letter giving details of the institute and instructions for completing the survey; a stamped addressed envelope for returning the survey and a packet of flower seeds as a small gift were also enclosed. The covering letter was signed by the Project Director. The 500 households were randomly chosen from a commercial phone directory database, Kurofune 2004 (Datascape & Communications Inc., 2004) for Toki City in the Gifu prefecture, and the Kita ward, Nagoya City in the Aichi prefecture. In all cases, the replies to the surveys were anonymous except that respondents gave their name and address if they wanted a summary of the survey report.

Fourth, a reminder postcard was sent to all recipients of the sample approximately two weeks after the initial mailing. The postcard thanked those who had already responded and requested a response from those who had not yet responded. Fifth, a new covering letter, also signed by the Project Director, a questionnaire, stamped addressed envelope, and packet of flower seeds were sent to those who said they had not received the questionnaire and would like to respond. As a result, of a total of 962 surveys that were validly distributed (479 in Kita ward, Nagoya City and 483 in Toki City, respectively), questionnaires from 428 households (201 from Kita ward and 227 from Toki City) were received by mail, for a response rate of 44.5%. Finally, a summary of the survey report, with a letter of thanks signed by the Project Director, was mailed to those who had requested it in late June, 2004.

The questionnaire included 26 questions, and 148 detailed items (Table 1). The questions concerned household characteristics, flood experience, risk perceptions, flood preparedness, willingness to pay to prevent flood damage, and choice experiments on flood prevention policies. To improve the response rate on important individual survey items that survey respondents are often unwilling to answer, such as household income and age, intervals were used rather than exact values. Respondents were asked to express WTP values for internal flood measures, external flood measures, and early-warning systems. Internal and external flood measures relate to structural modifications to waterways, while early-warning systems relate to non-structural measures. Because appropriate measures for external flood risk reduction differ between the two cities, this issue is discussed separately below.

3.3 Basic statistical results

Males accounted for 76.5% of all respondents. The range from 40 to 70 accounted for 71.6% of all respondents. Single-family houses accounted for 84.5% of the dwellings (compared to 57.5% for all of Japan in 1998). Wooden houses totaled 67.2% of the dwellings (53.5% for all of Japan in 1998), and houses with embankments composed 13.2% of the total.

The residence periods in the study area were distributed as follows: less

Table 1. Items in questionnaires.

Category		Item and definition
WTP	CVM	Internal floods: floods due to reduced sewage and pumping capacity during rains. External floods: floods due to collapse or overflow of dikes and/or dams as a result of rain exceeding that expected on the basis of probability Early-warning system: system for predicting an imminent flood and warning those in the area of risk
	Attributes of choice experiment	external flood reduction internal flood reduction early warning systems environmental protection willingness to pay for countermeasures
WTP factors	Residentsí attributes	Age: at 10 year intervals: 10s, 20s... Income: at 2 million yen intervals: less than 2 million, 2.01 - 4 million yen... Number of people in household: persons Occupation: Residence period: years Education: junior high school, high school, college school, university, graduate school
	Characteristics of house	Style: single family or multi-family Structure: wooden or non-wooden Ownership: rental or private Distance from river: 1) less than 100 m, 2) 100 - 500 m, 3) 500 m - 1 km, 4) 1 - 2 km, 5) 2 - 5 km, 6) more than 5 km
	Flood risk perception	Experience of flood disaster: yes or no Perceived frequency of flood risk: 0, 1,..., 6 corresponding to, once for 5 years, 10 years, 20 years, 50 years, 100 years, more than 100 years, and never respectively Perceived consequences of flood risk: 0 for not worried,..., 10 for very worried
	Perception of other risks	*The 25 items below were evaluated using 0 for not worried,..., 10 for very worried* Natural disasters: earthquakes, volcanoes, thunderstorms Environmental risks: pollution, global warming, endangered species Disease risks: cerebral apoplexy, cardiac insufficiency, AIDS, SARS, BSE Urban risks: gas explosions, fires, traffic accidents Traditional risks: labor accidents, robbery, murder High-technology risks: nuclear accidents, GMO, Internet damage
	Flood risk acceptability	Acceptability of above- and below-ground inundation
	Disaster preparedness	*The 18 items below were evaluated using yes (1) or no (0)* Insurance, evacuation kits, embankments, evacuation, familiarity with disaster maps, etc.
	Information provision	Effects of flood control works on environment: yes or no Local budget for public facilities like firefighting: yes or no
	Regional features	Upper reaches: Toki City Lower reaches: Kita ward of Nagoya City

than 10 years for 9.9% of the respondents, 10–20 years for 19.3%, 20–30 years for 21.1%, 30–40 years for 18.3%, 40–50 years for 12.2%, 50-60 years for 9.4%, and more than 60 years for 9.7%.

The distribution of annual income per household was as follows: 9.2% with less than 2 million yen, 24.9% with 2–4 million yen, 27.2% with 4–6 million yen, 13.7% with 6–8 million yen, 8.7% with 8–10 million yen, 5.9% with 10–12 million yen, 4.8% with 12–14 million yen, and 5.6% with more than 14 million yen.

The education levels of the respondents were distributed as follows: 18.5% had graduated from junior high school, 49.6% from high school, 11.7% from junior college, 18.5% from college, and 1.3% from graduate school.

4 Public Preferences for River Management

Respondents were asked to answer the question: "Although dams and dikes and so on are constructed to prevent flood occurrence, flooding occurs somewhere every year. What do you think of current flood prevention measures? Please choose your favorite ONE AND ONLY ONE measure from those below." Analysis of the results showed the diversity of people's interests in river management, although most (82%) thought that some flood control measures should be taken (Fig. 3). Respondents in Toki City had a greater preference for external flood measures (28.2%) than those in Kita ward (19%), while those in Kita ward had a greater preference for internal flood measures (48.1%) than those in Toki City (25.7%) (Fig. 4).

5 Flood Risk Acceptability

5.1 Perceived probability of house flooding in the future

In the survey, two questions dealt with residents' perceived a probability of their houses flooding in the future. The questions were worded as follows: "How often do you think your house will be flooded below (above) ground in the future? Please choose the most appropriate answer from the choices below. Once in 5 years, 10 years, 20 years, 50 years, 100 years, or more than 100 years; absolutely never; don't know." Because the recurrence of a 100-year flood or a 200-year flood may have been difficult for respondents to understand, the probability of a flood occurring within the next 50 years was also included in the survey. Of 391 and 364 valid responses to the questions of below- and above-ground inundation, respectively, the respective percentages of respondents answering "don't know" were 30% and 34%. If the "don't know" responses are excluded, the results (Fig. 5) indicate that nearly 60 and 70% of respondents did not correctly perceive the probabilities of below- and above-ground inundation, respectively. Among the group that did correctly perceive the probability of flooding, the median below-ground inundation probability was once in 50 years and once in 20 years (both 11.4%), while the median above-ground inundation probability was once in 100 years (9.6%).

5.2 Acceptable flood probability

Regarding the acceptable flood probability, the survey asked the following question: "What do you think about flood occurrence? Please choose the

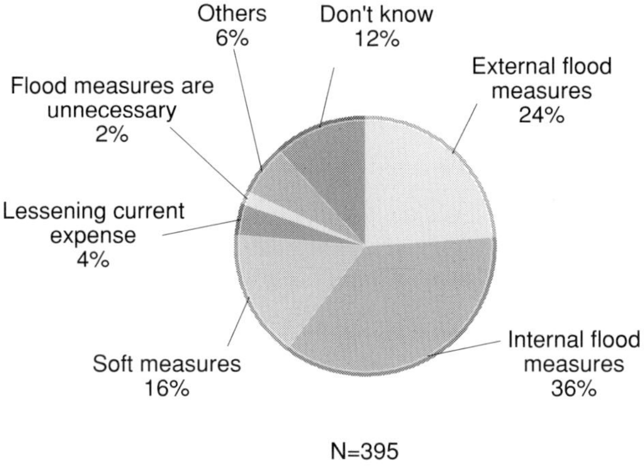

Fig. 3. Public preferences for river management.

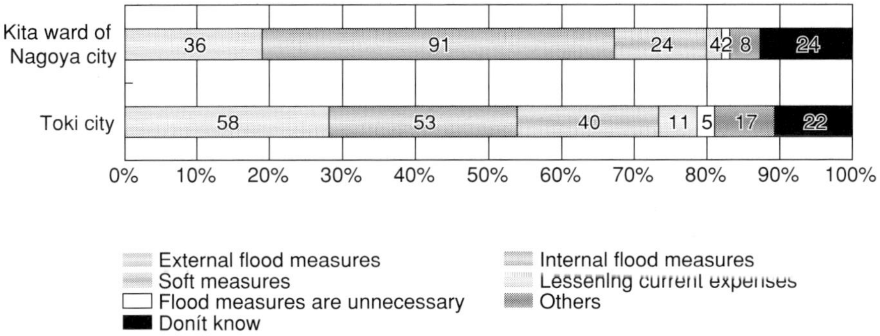

Fig. 4. Difference in public preferences according to regions.

most appropriate answer from the choices below. A. Although I live in a flood-prone area, I consider absolutely no flood occurrence acceptable. B. Because I live in a flood-prone area, I have no choice but to accept flood occurrence to some extent. C. Don't know." Of 312 valid responses, "absolutely unacceptable," "acceptable to some extent," and "don't know" constituted 37, 25, and 38% of the responses, respectively. Excluding the "don't know" group, more than half of the respondents accepted flooding to some extent. Of the "acceptable to some extent" group, residents accepting below- and above-ground inundation no more frequently than once in 100 years accounted for 88 and 77%, respectively (Fig. 6). The median below-ground inundation acceptability was once in 50 years (23%), while the median above-ground inundation

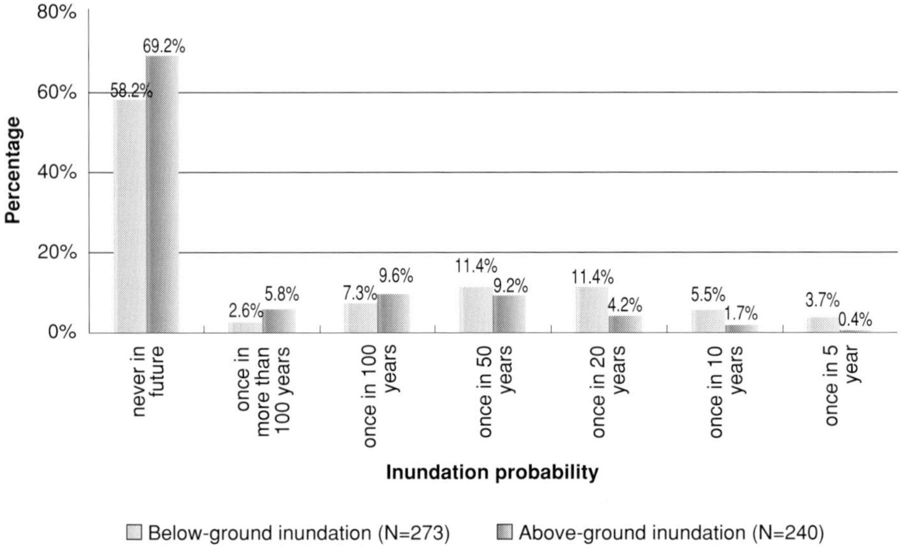

Fig. 5. Perceived probability of house inundation.

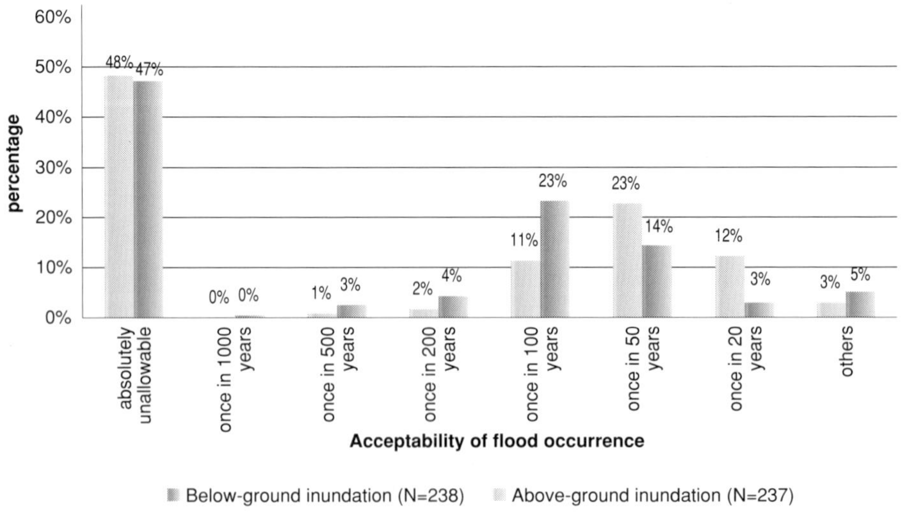

Fig. 6. Distribution of acceptable flood probability.

acceptability was once in 100 years (23%).

6 WTP for Flood Risk Reduction

As a matter of common sense, countermeasures against internal and external floods and early-warning systems should be implemented. These are defined in Table 1. First, respondents were asked to give answers for five scenarios: two scenarios each for internal and external floods, and one scenario for early warning systems (Table 2). The current sewage systems in Toki and Nagoya were designed for about a 5-year internal flood, while the levees and banks were designed for a 100-year flood in Toki and a 200-year flood in Nagoya. Planning flood control projects for higher levels requires an understanding of residents' WTP thresholds.

In addition, to examine WTP for environmental risks from flood control projects, information about potential damage to the river environment from the construction of flood control projects was either included in the survey or not. Here, environmental effects were dealt with as external factors; that is, when flood control projects are constructed, the environment may be damaged. However, Clark *et al.* (2002) treated them as internal factors; that is, when flood control projects are constructed, efforts might be made to minimize effects on the environment. The effect of providing financial information was considered in the same way.

Because the recurrence of floods at specific levels, i.e. 100-year or 200-year floods, may have been difficult for respondents to understand, the probability of a 50-year flood was also included in the survey in the same terms as once every 200 years (i.e., a probability of 22.2% within 50 years).

After indicating that the hypothetical projects would last for 20 years, respondents were asked to give their WTP values for each project. Because payment cards and dichotomous choice formats like a referendum approach are recommended and the former are more informative and cheaper to implement than the latter (Bateman *et al.*, 2002), payment cards were used in the survey.

The payment vehicle may affect WTP. There is some evidence that WTP in the form of tax is less than that in the form of donations (e.g., Andreoni, 1989; Champ *et al.*, 1997; Carson *et al.*, 1999; Chilton and Hutchinson, 1999). However, Hidano and Kato (2000) reported that the mean WTP for a tax format was higher than that for the donation format often used in Japan, because respondents could increase their utilities from just paying taxes; i.e., a "warm-glow of giving" effect existed in the tax format. Thus, to establish a more realistic WTP for flood control measures, even though they are typically financed

Table 2. Scenarios for flood control measures.

Region	Reaches	Scenarios for flood control measures			Number of samples (validly distributed)	Number of observations (response rate)
		Structural measures		Early warning systems		
		Internal floods (status quo: 5-year flood)	External floods (status quo: 100-year flood for Toki; 200-year flood for Kita ward, Nagoya City)	Fatality ratio (status quo: 1/10,000)		
Kita ward, Nagoya City	Lower reaches	10-year flood	500-year flood	1/20,000	500 (479)	201 (42%)
		20-year flood	1,000-year flood			
Toki City	Upper reaches	10-year flood	200-year flood		500 (483)	227 (47%)
		20-year flood	500-year flood			
Total					1000 (962)	428 (44.5%)

Q: If an early warning system for floods was established, the fatality ratio of a 50-year flood would decrease from one in every 10,000 people to one in 20,000. If a foundation was established for the specific purpose of implementing this early warning system, how much would you donate to it every year? (Choose one please)

1.	¥0	5.	¥2,000	9.	¥7,000	13.	¥25,000
2.	¥200	6.	¥3,000	10.	¥10,000	14.	¥30,000
3.	¥500	7.	¥4,000	11.	¥15,000	15.	¥35,000
4.	¥1,000	8.	¥5,000	12.	¥20,000	16. Other: ¥	

Fig. 7. Question asked in survey.

by the government, payment by donation was used in the survey (Fig. 7).

Because the status quo protection levels provided by levees in Nagoya (200-year flood) and Toki (100-year flood) cities differ, the WTP for external flood control measures for the two areas is discussed separately. Table 3 shows a statistical summary of the WTP for each scenario. The WTP levels for different measures range from ¥2,887 to ¥4,861 in terms of the mean and from ¥1,000 to ¥2,000 in terms of the median. Regarding the scenarios for flood control measures listed in the table, external scenario 1 refers to a measure that would improve the protection level from the status quo of a 100-year flood to that of a 200-year flood for Toki or from the status quo of a 200-year flood to that of a 500-year flood for Nagoya. External scenario 2 would improve the protection level from the status quo of a 100-year flood to that of a 500-year flood in Toki or from the status quo of a 200-year flood to that of a 1000 year flood for Nagoya. Internal scenario 1 refers to a measure that would improve the protection level from the status quo of a 5-year flood to that of a 10-year flood, while internal scenario 2 would improve the protection level to that of a 20-year flood.

6.1 Determinants of WTP for flood risk reduction with multivariate regression

Items in Fig. 1 differed in terms of indices and data sufficiency in the questionnaire. Some items, like limited resources, had only one index, while others, like individual preparedness, perceived flood probability, and consequently, had several indices. Therefore, it was necessary to decide which method to use, and which dependent variables to construct to test the hypothesis for the theoretical framework in Fig. 1. Here, the framework was first tested by multivariate regression, the dependent variables of which were the WTP for each measure for reducing flood risk with the independent variables being flood experience, flood perception (probability and consequences), perception of other risks, acceptability of flood risk, limited resources (income

Table 3. Statistical summary of WTP for each scenario (¥ per respondent per year).

			Mean	95% confidence interval of mean	Median
Scenarios for flood control measures	External flood control measure scenario 1	Kita ward, Nagoya City (from status quo 200-year flood to 500-year flood)	3,932	2,838 - 5,027	2,000
		Toki City (from status quo 100-year flood to 200-year flood)	4,665	3,516 - 5,815	2,000
	External flood control measure scenario 2	Kita ward, Nagoya City (from status quo 200-year flood to 1000-year flood)	3,674	2,438 - 4,910	1,000
		Toki City (from status quo 100-year flood to 500-year flood)	4,861	3,573 - 6,150	2,000
	Non-structural measures (from status quo 1/10,000 to 1/20,000)		2,887	2,346 - 3,418	1,000
	Internal flood control measure scenario 1 (from status quo 5-year flood to 10-year flood)		2,927	2,409 - 3,443	1,000
	Internal flood control measure scenario 2 (from status quo 5-year flood to 10-year flood)		3,152	2,557 - 3,748	1,000

per capita), individual preparedness, etc.

Due to high correlations between perceived probability of under- and above-floor inundation (0.914), the perceived under-floor inundation probability was selected as one of the independent variables (PP) of WTP. Individual preparedness (IP) was assessed from 18 sub-items, like preparation of survival food, potable water, hazard map, and so on. If the answer was YES to the preparation of a sub-item, then 1 mark was given. If all answers were YES, the individual preparedness mark was 18. The degree of worry about other risks (WO) and perceptions of the consequences of flood risk (CP) were averaged by degrees of worry ranging from 0 (not at all) to 10 (very worried) from 25 non-flood-related risks (excluding typhoons, storms, river flooding, and landslides) and 12 sub-items of flood impacts, respectively. Income per capita (INCO) was used to represent limited resources. In addition, flood experience (EXPE), the effects of providing environmental and budget information (PEI and PBI) on WTP, and the distance from a river (DIST) were tested in the models. A multivariate regression model was run with SPSS 10.0J for Windows (SPSS Inc., 1999a). The results are shown in Table 4.

First, the goodness of fit to the models was 0.40–0.45 for external measures, 0.22–0.23 for internal measures, and 0.126 for non-structural measures. This suggests that the framework in Fig. 1 is most applicable to external measures.

Second, each variable did not necessarily play the same role in each model,

Table 4. Results of multivariate regression analysis (Significance probability in parentheses).

Variables	Assumed signs in models	Scenario 1 for Kita ward, Nagoya City	Scenario 1 for Toki City	Scenario 2 for Kita ward, Nagoya City	Scenario 2 for Toki City	Non-structural measures	Internal flood control measure scenario 1	Internal flood control measure scenario 2
Constant	±	88.5 (0.988)	16520.6 (0.218)	522.2 (0.940)	-4685.1 (0.791)	460.0 (0.870)	4353.1 (0.165)	5594.4 (0.156)
IP: Individual preparedness	+	287.5 (0.397)	438.5 (0.480)	477.3 (0.275)	347.7 (0.591)	76.2 (0.666)	311.5 (0.106)	286.8 (0.224)
INCO: Income per capita	+	5.3 (0.455)	11.8 (0.393)	2.3 (0.801)	20.1 (0.174)	3.0 (0.406)	**7.0** (0.072)**	**8.2** (0.089)**
PP: Perceived probability of under-floor inundation	+	-15.0 (0.977)	-393.8 (0.668)	293.8 (0.648)	78.9 (0.934)	-123.9 (0.597)	23.8 (0.928)	120.1 (0.727)
FA: Flood risk acceptability	-	603.2 (0.169)	620.7 (0.264)	429.5 (0.452)	9.7 (0.986)	160.9 (0.358)	5.7 (0.978)	-65.6 (0.799)
CP: Perceived consequences of flood	+	-1013.3 (0.162)	-1196.1 (0.401)	-501.9 (0.530)	424.1 (0.818)	13.6 (0.956)	-219.5 (0.420)	-254.9 (0.459)
WO: Worry about other risks	-	1018.3 (0.175)	-341.5 (0.738)	436.1 (0.583)	829.6 (0.427)	**540.1** (0.058)**	150.8 (0.661)	16.8 (0.968)
EXPE: Flood experience	+	**5309.1* (0.046)**	-4715.5 (0.171)	**6318.1* (0.039)**	-1247.3 (0.747)	29.4 (0.976)	-780.7 (0.474)	369.9 (0.783)
PEI: Providing environmental information	-	-2882.2 (0.165)	-2857.0 (0.307)	**-5162.2** (0.070)**	-3675.7 (0.191)	-956.0 (0.260)	**-2779* (0.004)**	**-4088.6* (0.001)**
PBI: Providing budget information	±	1319.9 (0.504)	1361.1 (0.720)	-1782.3 (0.500)	-1019.8 (0.793)	298.7 (0.738)	417.6 (0.671)	-1035.6 (0.399)
DIST: Distance from river	-	**-1783.6* (0.038)**	887.7 (0.481)	**-2123.4* (0.022)**	-137.6 (0.918)	**-586.2* (0.041)**	-378.4 (0.273)	-532.8 (0.231)
R -squared		**0.446**	**0.456**	**0.405**	**0.423**	**0.126**	**0.219**	**0.235**

Note: * and ** denote statistically significant levels of 0.05 and 0.01, respectively.

and so was not necessarily consistent with the assumptions in the framework. IP, INCO, and PEI had the same signs in the seven models, while the other factors did not. This suggests that WTP may increase with individual preparedness and income per capita, but may decrease after negative information on flood control measures is provided, which is consistent with the common assumptions. Of the variables with different signs in the models (PP, FA, CP, WO, EXPE, PBI, and DIST), some were interrelated—for example, flood risk acceptability depends on the trade-off between flood risk perception and the perception of other risks. This implies that another method (e.g., a covariance structure analysis) is needed to obtain a general and integrated evaluation.

Finally, the main factors with a statistically significant level of 0.1, all for Kita ward, Nagoya City, were EXPE and DIST for external measures;

EXPE, PEI, and DIST for internal measures; WO and DIST for non-structural measures; and INCO and PEI for internal measures. However, there were no statistically significant factors for either external or internal measures for Toki city.

6.2 Covariance structure analysis

As mentioned, the multivariate regression analysis showed slightly different results for the determinants of individual WTP for each flood risk reduction measure. To discuss the general determinants of a synthesized WTP for flood risk reduction, which are measured indirectly by means of observable indicators, such as individual WTP, a covariance structure analysis (CSA) was used here.

CSA is an extension of the regression model and is used to test the fit of a correlation matrix against two or more causal models being compared (e.g., Krzanowski and Marriott, 1998; Wakui and Wakui, 2003). CSA originated from physiology, and has been widely applied in business, marketing, and human resource management. In CSA, the researcher uses the existing knowledge/theory/framework to generate hypotheses of how the system functions. These hypotheses are explicitly stated in the form of a causal model that depicts pathways, both direct and indirect, by which (latent or observed) variables influence each other. The researcher's theoretical model can be evaluated by assessing the extent to which covariances among variables in the model are consistent with those occurring in the actual data. To test the theoretical framework shown in Fig. 1 through CSA, two latent variables, general WTP and perceived flood risk, were introduced. The general WTP represents the resident's fundamental attitude to paying for a reduction in flood risk and determines the specific WTP for each measure, while the perceived flood risk is conceptualized from perceptions regarding the probability and consequences of flood risk.

Amos 4.0 for Windows (SPSS Inc., 1999b) was used to conduct the covariance structure analysis of WTP for flood control measures, and the results of the direct effects for each factor in WTP are shown in Fig. 8. Here, each effect is denoted by an arrow with the tail at the cause and the head pointing to its direct effect. A direct effect is represented by a single arrow, whereas indirect effects involve paths of two or more linked direct effects. Spurious effects (non-causal correlations) between variables of flood risk perception and perception of other risks are indicated by double-headed arrows. The values above the line show the degree of direct effects or non-causal correlation of variables, while those in parentheses below are the significant probabilities of

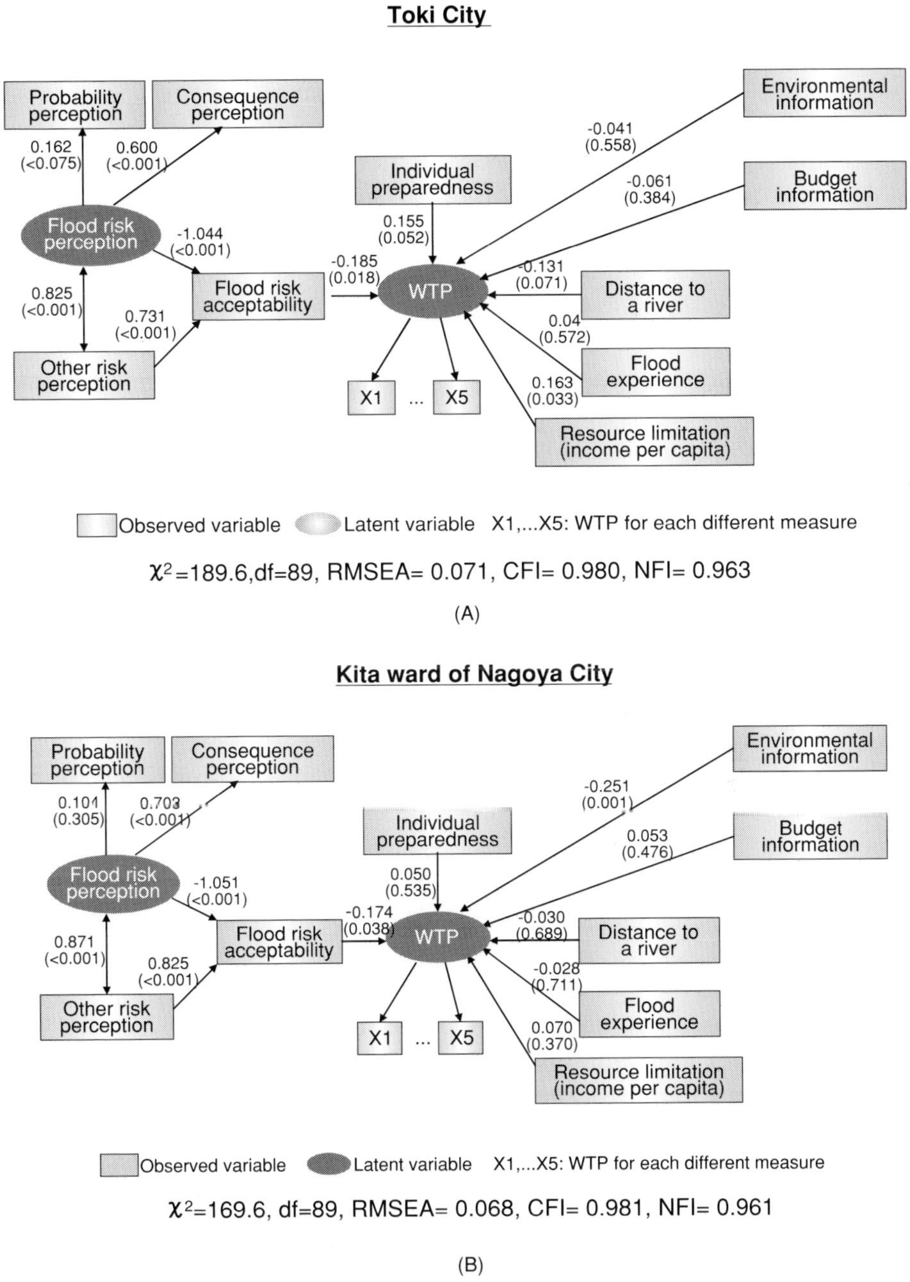

Fig. 8. Simplified path diagrams of WTP for flood control measures.
Note: Covariances and error terms associated with endogenous variables have been omitted from the diagrams for clarity.

the effects (standardized regression weights) or the correlations.

First, the proposed models for both Toki City and Kita ward, Nagoya City appear to be supported. The root mean square error of approximation (RMSEA), comparative fit index (CFI), and normed fit index (NFI) are approximately 0.070, 0.98, and 0.96, respectively. In addition, the chi-square to degree of freedom ratio ranges from 1.91 to 2.13. Together, these fitting statistics suggest that the model fits the data reasonably well[1].

Second, the signs for the effect values of the variables are consistent with the assumption in Table 4, implying the validity of the framework in Fig. 1. That is, in relation to the direct effects of the variables, WTP for flood control measures may increase with income per capita, individual preparedness, and flood experience, but may decrease with distance to a river, flood risk acceptability, and the provision of environmental information. Providing budget information with different signs showed different impacts on WTP in Toki City and Kita ward, Nagoya City. However, in relation to indirect effects, the signs of the variables imply that WTP increases with perceived flood risk and decreases with perception of other risks.

Finally, the most important determinants of WTP for flood control measures are slightly different for both areas. For Toki City, the most important determinants were acceptability of flood risk, individual preparedness, income per capita, and the distance to a river, while for Kita ward, they were acceptability of flood risk and the provision of environmental information, at a statistically significant level of 0.1. Furthermore, the acceptability of flood risk depends (at a statistically significant level) on perceived flood risk and perception of other risks. The effect values suggest that perceived flood risk plays a slightly more important role in the acceptability of flood risk than perception of other risks.

7 Multi-Attribute Evaluation with Choice Experiment Approach

7.1 Framework for evaluating multi-attribute flood prevention measures through a choice experiment

In a choice experiment, respondents are presented with a series of alternatives and asked to choose those that they most prefer. A baseline alternative, corresponding to the status quo, is generally included in each choice set. Usually, each alternative is defined by a number of attributes, which vary among the different alternatives. Aggregated choice frequencies are modeled to infer

[1]The magnitudes of the fitting statistics should be interpreted in light of the fact that individual items were analyzed rather than multi-item composites, which would more closely satisfy the assumption of multivariate normality.

the relative impact of each attribute on choice, and the marginal value of each attribute for a given option is calculated by statistical methods like the multinomial logit model. Along with the attributes, individual characteristics such as income and age may also influence the choice. The mixed logit model can be used to deal with these characteristics (Greene, 2003).

The main theoretical support for the choice experiment technique is the random utility theory (Thurstone, 1927; Mcfadden, 1973; Manski, 1977), according to which consumers maximize their utility function (subject to a budget constraint), whose random term is supposed to have a specific distribution:

$$u_i = v_i + \varepsilon_i, \tag{1}$$

where U_i is the utility when the ith scenario is chosen, V_i is the deterministic component, and ε is the random term.

Supposing that the random terms have an extreme-value (Gumbel) distribution, the probability of choosing the ith scenario from a choice set Y follows a logistic distribution and leads to what is called the conditional logit model (Mcfadden, 1973; Greene, 2003):

$$P(i/Y) = \exp(\lambda V_i)/\sum_j \exp(\lambda V_i). \tag{2}$$

An important implication of the standard logit model is that selections from the choice set must obey the property of independence from irrelevant alternatives (IIA), which states that the relative probabilities of two options being selected are unaffected by the introduction or removal of other alternatives (Luce, 1959).

To estimate the indirect utility function, the following linear form is often applied:

$$V_{in} = A_i + \sum_j \beta_j x_{ij} + \sum_h \alpha_h z_{hn}, \tag{3}$$

where A_i is an alternative-specific constant (ASC), β_j is the parameter of the jth attribute of the ith alternative represented by the variable x_{ij}, and α_h is the parameter of the hth characteristic of person n represented by z_{hn}.

As a measure of the benefits resulting from changes in an attribute, the marginal willingness to pay (MWTP), which is widely discussed as a main research consideration in the fields of transportation and environmental studies, can be rewritten as

$$\text{MSR}_i = -\frac{\partial V/\partial x_i}{\partial V/\partial \text{ price}}. \tag{4}$$

The price level used here was the mean payment increase proposed in each survey scenario, while V is the marginal indirect utility for attribute i.

In a manner similar to MWTP, another measure, the marginal substitution ratio (MSR), provides the marginal substitution of one attribute due to a change in another one. Because of the specific functional form of the indirect utility function, MSRs have to be calculated as

$$\mathrm{MSR}_{j->i} = -\frac{\partial V/\partial x_i}{\partial V/\partial x_j}. \tag{5}$$

The implementation of the choice experiment approach proceeds in several steps, as described in a later section. We can already specify that flood prevention measures must first be described by their main attributes and can take different respective levels. The attributes and levels are then combined (using a statistical design) to create scenarios, which are formed into choice sets that are presented to respondents at the time of the survey.

7.2 Valuation question card

As a matter of common sense, countermeasures against internal and external floods and early warning systems should be implemented to prevent flood damage, yet at the same time, the environment should be protected at the lowest possible cost. Therefore, we utilized five criteria (attributes) comprising different public preferences for flood prevention measures: external flood reduction, internal flood reduction, early warning systems, environmental protection, and willingness to pay for countermeasures. These criteria were represented as the occurrence of disastrous floods, inundation depth, fatality rate, environmental improvement ratio, and annual additional expense per capita, respectively. Four levels were assigned to each attribute (Table 5).

To help individuals understand each attribute, brief explanations were provided in the survey questionnaires. The chance of disastrous flood occurrence was defined as the probability of floods due to the collapse or overflow of dikes and dams (i.e., an external flood) as a result of rains exceeding those that would be expected on the basis of probability. At the moment, the dikes in Nagoya city are built to withstand 200-year rains, while those in Toki city are built to withstand 100-year rains. The levels for this attribute were defined as −10, −20, −50%, and the status quo (0%). Inundation depth reduction was defined as the decrease in the depth of inundation due to reduced sewage and pumping capacity during rains (i.e., an internal flood). The levels for this attribute were defined as −10, −20, −50 cm, and the status quo (0 cm). The fatality rate due to floods was defined as the number of deaths due to

Table 5. Attributes and levels in the choice experiment.

		Attribute levels			
Attributes		Status quo	1	2	3
Measures/ goals	Chance of reducing disastrous flood occurrence (%)	0	-10%	-20%	-50%
	Inundation depth decrease (cm)	0	-20	-50	-100
	Fatality rate due to floods	1 in 10,000	0.9 in 10,000	0.8 in 10,000	0.5 in 10,000
	Improvement rate for river environment	0	10%	20%	50%
Annual additional expense per capita		¥0	¥2,000	¥5,000	¥10,000

a flood over the total population affected by the flood. The current fatality rate due to floods is approximately 1 in 10,000 per flood event (Zhai *et al.*, 2006a). Here, the levels were specified as 0.9 in 10,000, 0.8 in 10,000, 0.5 in 10,000, and the status quo (1 in 10,000). Because the occurrence of fatalities largely depends on early warning and evacuation systems, the fatality rate reduction may be viewed as its representative index. The river environment refers to the natural environment, including the aquatic plants and animals in a river and in the waterfront ecosystem, and the living environment at the interface between the river and human beings, including the water quality, landscape, and river space. The levels for this attribute were defined as improvement rates of 10, 20, and 50%, and the status quo (0%). The last attribute was the additional expense of flood prevention measures, with levels of 2,000 yen/person/year, 5,000 yen/person/year, 10,000 yen/person/year, and the status quo (0 yen/person/year). Based on the attributes and their levels, 12 choice sets were created by an orthogonal design approach using SPSS version 10.0J (SPSS Inc., 1999a). The valuation section of the survey consisted of four separate questions. For each question, respondents were asked to choose the most desirable of three alternatives based on descriptions of flood prevention measures at different prices (options A, B, and C), or to choose option D (the status quo) (Fig. 9).

7.3 Main results

To find detailed relationships between the utility and the attributes or individual characteristics, the analysis results can be discussed in terms of four aspects. The first is to provide a statistical summary to help understand the results. The second is to evaluate the two types of model that we applied—one containing only attributes and the other containing both attributes and individual characteristics—to see whether they correctly describe the data and to determine which variables significantly affect each model. The third is to ob-

Please examine each question below and choose ONE AND ONLY ONE option

	Option A	Option B	Option C	Option D (status quo)
Chance of disaster occurrence	10% decrease	status quo	20% decrease	status quo
Inundation depth	status quo	20 cm decrease	100 cm decrease	status quo
Fatality rate due to floods	0.8 in 10,000	0.9 in 10,000	status quo	1 in 10,000
Improvement rate for river environment	status quo	50% improvement	status quo	status quo
Annual additional expense per capita	¥5,000	¥5,000	¥5,000	¥0
	↓	↓	↓	↓
I would select	A.	B.	C.	D.

Fig. 9. Valuation question card from the questionnaire.

tain important information like the marginal substitution ratio (MSR) from the validated model or models. The fourth is to discuss whether there are regional differences in the model results.

Table 6 shows the results for two multinomial logit models, denoted as Model 1 and Model 2, which were both processed with LIMDEP Version 8.0 (Greene, 2002). Model 1 contained the attributes and constants, while Model 2 was a full model containing both the attributes and the socio-economic factors. To correctly draw inferences from the model results, it was important to validate the models. The validation included evaluating the goodness of fit and assessing each variable coefficient, including the alternative-specific constants (ASCs). The explanatory power of Model 2 was a little stronger, with an R-squared and an adjusted R-squared of 0.232 and 0.221, respectively.

The hypothesis that the socio-economic factors are not significant in explaining the respondents' choice preferences was strongly rejected by the likelihood ratio test (Eq. (6)): chi-squared = 2[852.174 − 822.13] = 60.1. The critical chi-squared value for 24 degrees of freedom is 36.42. This means that the socio-economic factors significantly affected an individual's decision-making, and thus, Model 2 was statistically proven to be stronger than Model 1 in terms of its explanatory power. The multinomial logit model necessarily involves strong assumptions about the independence of irrelevant alternatives (IIA). Hausman tests for IIA reject the null hypothesis.

$$\begin{aligned}\text{Chi-squared} &= 2(\ln L(\text{model with a factor}) \\ &\quad - \ln L(\text{model without a factor})) \\ &> \text{Chi-squared (degrees of freedom).}\end{aligned} \tag{6}$$

Four attributes (CHANCE, DEPTH, ENVIRONMENT, and EXPENSE) were statistically significant, and their coefficient signs were the same in both models, while the attribute of FATALITY was statistically insignificant in both models. The large *p*-values of the FATALITY variable in both models (0.97 and 0.88) imply that this attribute was almost totally ignored in decision-making on flood prevention measures when the ASCs and socio-economic factors were considered in the model.

All ASC coefficients were positive and statistically significant across all options in Model 1, but not in Model 2. ASCs can capture a mixture of status quo bias effects and the impacts of unobserved attributes in an attribute-only model (Bateman *et al.*, 2002). Thus, the different forms of the ASCs in Models 1 and 2 implied that the impact of unobserved attributes like the socio-economic factors was important, which is consistent with the model

Table 6. Results for two multinomial logit models with choice as a dependent variable.

	Model 1 (attributes and constants)	**Model 2** (full model)
CHANCE	**-0.997 (-3.52) ******	**-0.930 (-3.30)******
DEPTH	**-0.00632 (-3.37) ******	**-0.006 (-3.30)******
FATALITY	-0.00964 (-0.03)	0.044 (0.15)
ENVIRONMENT	**0.89 (3.06) *****	**0.910 (3.10)*****
EXPENSE	**-0.000215 (-8.37) ******	**-0.00022 (-8.40)******
ASCA	**1.82 (7.15) ******	-1.623 (-0.91)
ASCA x REGION		0.354 (0.83)
ASCA x INFOENV		-0.590 (-1.60)
ASCA x INFOBUDG		0.480 (1.30)
ASCA x EXPERIENCE		-0.540 (-1.30)
ASCA x SEX		**1.100 (2.0)***
ASCA x AGE		0.110 (0.65)
ASCA x INCOME		**0.330 (2.40)***
ASCA x EDUCATION		0.260 (1.20)
ASCA x DISTANCE		0.150 (1.30)
ASCB	**2.23 (8.92) ******	-0.2747 (-0.1617)
ASCB x REGION		0.172 (0.4176)
ASCB x INFOENV		-0.590 (-1.70)
ASCB x INFOBUDG		**0.720 (2.0)***
ASCB x EXPERIENCE		-0.620 (-1.60)
ASCB x SEX		**1.100 (2.10)***
ASCB x AGE		-0.096 (-0.58)
ASCB x INCOME		**0.270 (2.0)***
ASCB x EDUCATION		0.390 (1.80)
ASCB x DISTANCE		**0.300 (2.50)***
ASCC	**1.94 (6.39) ******	-1.4123 (-0.8007)
ASCC x REGION		0.266 (0.624)
ASCC x INFOENV		-0.560 (-1.50)
ASCC x INFOBUDG		0.430 (1.10)
ASCC x EXPERIENCE		**-0.980 (-2.40)***
ASCC x SEX		**1.300 (2.50)***
ASCC x AGE		0.087 (0.50)
ASCC x INCOME		**0.390 (2.90)*****
ASCC x EDUCATION		**0.430 (2.0)***
ASCC x DISTANCE		0.160 (1.30)
No. of observations	772	772
No. of choices	3088	3088
Log likelihood	-852.176	-821.59
R-sqrd	0.204	0.232
RsqAdj	0.201	0.221
Chi-squared	145.427	206.62
Prob[chi squared > value]	<0.00001	<0.00001

Note: 1) *t* value in parenthesis
2) *, **, ***, and **** refer to significance levels of 0.05, 0.01, 0.005, and 0.001, respectively.
3) ASCA, ASCB, and ASCC are the ASCs of Options A, B, and C with respect to Option D (status quo).

evaluation results mentioned above.

The effects of the socio-economic factors can be discussed in terms of their interaction with the ASCs, though the interpretation is complicated. First, the signs of all coefficients for the interaction terms with the socioeconomic variables, except for the AGE variable, were consistent with common knowledge. Second, the interaction terms with SEX and INCOME produced statistically significantly positive impacts on the utility, but the coefficients of the interaction terms with REGION, INFOENV, and AGE were statistically insignificant for all three non-status-quo options at the 0.05 level. Third, EXPERIENCE, INFOBUDG, EDUCATION, and DISTANCE were not significant for all three non-status-quo options at the 0.05 level. Here, we use the likelihood ratio test to discuss the effects of these variables on the models, based on the hypothesis that a certain factor is not significant in explaining the choice preferences for flood prevention measures. If the hypothesis is rejected by the likelihood ratio test (Eq. (6)), a factor's effect is significant. The chi-squared values and the probabilities given by the likelihood ratio test for certain factors listed in Table 7 show that EXPERIENCE and DISTANCE were statistically significant, while EDUCATION and IFORBUDG were not, at the 0.05 level.

7.4 Implicit relationships between attributes from model results

Table 8 shows the matrix of MSRs obtained from the results for Model 2 listed in Table 7. The variable FATALITY, which had little impact on decision-making (i.e., a high p-value), is marked with an asterisk. For Model 2, to maintain the same utility if other conditions do not change, a 10% improvement in the river ENVIRONMENT should be proportional to a 9.8% increase in the flood occurrence probability, to a 15.167 cm increase in the inundation depth, or to 413.6 yen per capita of additional annual expense. The last column in Table 8 lists point estimates of the marginal willingness to pay (MWTP) for each attribute.

Table 9 lists the implicit prices (MWTPs; in yen/person/year) and 95% confidence intervals for Models 1 and 2, as obtained by Monte-Carlo simulation with 1,000 iterations. The interval estimates show nearly the same values for each attribute, except for FATALITY, for both models. The similar results for MWTP imply that the choice experiment approach produced stable estimates that were nearly free of the socio-economic factors, but were affected by the ASCs. The fact that the confidence intervals for the attribute FATALITY crossed zero and had very small means in contrast to the deviations also implied that the MWTP for FATALITY may have been zero. Therefore, for Model 2, the MWTP values were 422.7 yen/person/year for a 10% decrease

Table 7. Chi-squared values from the likelihood ratio test for certain factors.

	Model 2
EXPERIENCE	8.24 (0.041)*
IFORBUDG	5.62 (0.132)
EDUCATION	5.60 (0.133)
DISTANCE	11.78 (0.008)**
REGION	1.1(0.777)

1) Probability in parentheses
2) *, **, ***, and **** refer to significance levels of 0.05, 0.01, 0.005, and 0.001, respectively.

Table 8. Marginal substitution ratios among the attributes in Model 2.

	CHANCE (100%)	DEPTH (cm)	FATALITY (/10,000 persons)*	ENVIRONMENT (100%)	EXPENSE (MWTP) (yen/person/year)
CHANCE (100%)	-1	-155	21.14	1.022	-4,227
DEPTH (cm)	-0.00645	-1	0.136	0.0066	-27
FATALITY (/10,000)*	0.04731	7.33	-1	-0.0484	200
ENVIRONMENT (100%)	0.97849	151.67	-20.68	-1	4,136
EXPENSE (yen/person/year)	-0.00024	-0.037	0.005	0.0002	-1

* Statistically insignificant at 0.1 level

Table 9. Implicit prices (MWTPs; yen/person/year) and 95% confidence intervals for each model by Monte-Carlo simulation with 1,000 iterations.

	Model 1	Model 2
CHANCE (100%)	-4,637 (-8,044 to -1,948)	-4,227 (-7,366 to -1,621)
DEPTH (cm)	-29 (-48 to -11)	-27 (-49 to -9)
FATALITY (/10,000)	-45 (-2,624 to 2,549)*	200 (-2,936 to 2,324)*
ENVIRONMENT (100%)	4,140 (1,387 to 7,098)	4,136 (1,458 to 7,157)

* Statistically insignificant at 0.1 level

in the flood occurrence probability, 270 yen/person/year for a 10 cm reduction in the inundation depth, and 422.7 yen/person/year for a 10% improvement in the river environment.

7.5 Result differences between regions

The fact that the variable REGION in the model proved insignificant according to the pooled (total) data implied that REGION statistically had no significant impact on the utility functions of Kita ward and Toki City. This does not necessarily mean, however, that the analysis results would have been the same in each region had we used regionally separated data. First, the discussion with regard to the likelihood ratio test concerns the involvement of the various socio-economic variables in the model and the results evaluated for different regions. In a manner similar to the above analysis, the former was discussed in terms of Eq. (6), while the latter was examined with Eq. (7) (Greene, 2003). This equation tests the alternative hypothesis that the constant term and the coefficients of the attribute variables were the same whether REGION equaled 1 (Toki City) or 2 (Kita ward). As a result, four cases were examined, and the results are listed in Table 10.

$$\begin{aligned}\text{Chi-squared} &= 2(-\ln L(\text{pooled}) + (\ln L(\text{model with a factor}) \\ &\quad + \ln L(\text{model without a factor}))) \\ &> \text{Chi-squared (degrees of freedom)}.\end{aligned} \tag{7}$$

The likelihood ratio tests on the model differences in the two regions showed that both models were significant at the 0.05 level. The tests on the differences in the two models, however, showed that the socio-economic factors were statistically significant in affecting the decision-making in Toki City (less than 0.0001), but not in Kita ward (0.1086), at the 0.05 level.

In addition, the differences were examined in terms of the coefficients of the attribute variables in the four models, as listed in Table 11. At least three things can be observed from the table. First, there were great differences not only in the coefficients, but also in the t-values of two variables, CHANCE and ENVIRONMENT, with respect to REGION. In the Toki City model, ENVIRONMENT was significant but CHANCE was not, while the opposite was true for Kita ward. Second, FATALITY was not statistically significant in any of the four models, which was consistent with the analysis results described above. Third, the coefficients for DEPTH and EXPENSE were nearly the same in all four models.

Finally, the degree of regional difference can be explained in terms of

Table 10. Significant probability for likelihood ratio tests between regions and models.

	Between regions	Toki city	Kita ward of Nagoya city
Between models 1 and 2		0.0001	0.1086
Model 1 (attributes and constants)	0.03373		
Model 2 (full model)	0.04449		

Table 11. Coefficients and t-values of the attribute variables in the four models.

	Kita ward of Nagoya city		Toki city	
	Model 1 (attributes and constants)	**Model 2** (full model)	**Model 1** (attributes and constants)	**Model 2** (full model)
CHANCE	-1.2753 (-3.121)	-1.15527 (-2.781)	-0.71314* (-1.809)	-0.65392* (-1.625)
DEPTH	-0.0061 (-2.255)	-0.00631 (-2.286)	-0.00626 (-2.395)	-0.00639 (-2.404)
FATALITY	0.4085* (0.992)	0.54369* (1.258)	-0.40976* (-1.058)	-0.42590* (-1.063)
ENVIRONMENT	0.4814* (1.139)	0.58882* (1.353)	1.25911 (3.112)	1.35282 (3.227)
EXPENSE	-0.00025 (-6.471)	-0.00026 (-6.476)	-0.00019 (-5.368)	-0.00020 (-5.444)

1) t-value in parentheses
2) * Statistically insignificant at 0.05 level

Table 12. Regional differences in MWTP (yen/person/year) for Models 1 and 2.

	Model 1 (attributes and constants)	**Model 2** (full model)
CHANCE (100%)	-1,197	-1,381
DEPTH (cm)	7.7	8.3
FATALITY (/10,000)	4,284	3,833
ENVIRONMENT (100%)	-4,560	-4,738

MWTP instead of MSR because of the complexities of the MSR matrices. Table 12 shows slight differences in the MWTPs between Toki City and Kita ward for both models. In the case of Model 2, as compared with those of Toki City, the MWTPs for Kita ward were 138.1 yen/person/year more for a 10% decrease in the flood occurrence probability, 83 yen/person/year less for a 10 cm reduction in the flood inundation depth, 383.3 yen/person/year less for a fatality reduction of 1/100,000, and 473.8 yen/person/year less for a 10% improvement in the river environment. Therefore, it seems that Nagoya City's Kita ward pays more attention to disastrous flood occurrences, while Toki City focuses more on environmental protection and fatality reduction measures.

8 Concluding Remarks

This study investigated public preferences regarding flood control measures, WTP, the main factors determining WTP, and the relationships among five attributes of flood risk reduction measured through a survey conducted in the Shonai-Toki River region of Japan from the end of March to the beginning of April, 2004. The main findings are summarized as follows.

- Most residents expect some flood control measures and have diverse interests in river management.
- Nearly half of the respondents accept no flood risk at all.
- The WTP levels for different measures range from ¥2,887 to ¥4,861 in terms of the mean and from ¥1,000 to ¥2,000 in terms of the median. It is likely that there is zero marginal WTP for flood risk reduction in both Toki City and Nagoya City. This is because WTP for flood risk reduction must be determined within a multi-risk context.
- Among the five attributes of the flood prevention measures, four attributes—specifically, the chance of reducing the number of disastrous flood occurrences, the reduction of inundation depth, the rate of improving the river environment, and the additional expense of new flood prevention measures—were statistically significant.
- There was a statistically significant difference in the model results between Toki City and Kita ward of Nagoya City.

How to apply CVM and the choice experiment approach the general cautions have been broadly discussed in environmental economics (e.g., Bateman, 2002, p. 74). However, when they are applied to flood risk reduction, additional cautions are necessary.

First, WTP for flood risk reduction should be considered together with the risk curve. That is, the "scope effect" in economics may not be detected at a statistically significant level due to its very small changes in WTP. Flood risk, particularly in terms of fatality, is very small, usually at a level of less than 10^{-6}. In the case of Japan, the number of annual fatalities due to floods is usually less than 100. People may not take it very seriously because they perceive that other risks, such as fire or earthquake risk, are more important than flood risk. This is reflected in the WTP where their marginal WTP may be close to zero. Actually, results for fatality reduction in both CVM and the choice experiment were statistically insignificant when socio-economic factors were included in the analyses. The value of statistical life (VOSL) obtained by other methods, such as a hedonic approach, should be used to calculate the benefit of fatality reduction due to flood risk reduction measures.

Second, the results for model analysis contain uncertainty. As in the analysis for the choice experiment, the fatality factor is statistically significant when only attributes are input in the model, but becomes non-significant when additional socio-economic factors are included. It is important to clarify information regarding how the WTP is produced; for example, by checking what kind of question was asked, what kind of model was used, and how well the model performed.

Third and finally, WTP for moral satisfaction may be induced when CVM is used. Because the marginal WTP for flood risk reduction is close to zero, it would be better to understand that the estimated WTP may be regarded as the one for moral satisfaction.

References

Andreoni, J. (1989): Giving with Impure Altruism: Applications to Charity and Ricardian Equivalence, *Journal of Political Economy*, **97**(6), 1447–1458.

Arrow, K. J., P. R. Solow, E. E. Portney, R. Leamer, R. Radner, and H. Schuman (1993): Report of the NOAA panel on contingent valuation, *Federal Register*, **58**, 4601 (Jan. 15).

Bateman, L. J. and K. G. Willis (2001): Valuing Environmental Preferences, Oxford University Press.

Bateman *et al.* (2002): Economic Valuation with Stated Preference Techniques: A Manual, Edward Elgar Publishing Limited.

Carson, R. T. and R. C. Mitchell (1989): Using Surveys to Value Public Goods: The Contingent Valuation Method, Resources for the Future, Washington, DC.

Carson, R. T., R. C. Flores, and R. C. Mitchell (1999): The theory and measurement of passive-use value, I. Bateman and K. G. Willis eds., Valuing Environmental Preferences, Oxford University Press, pp. 109–123.

Champ, P. A., R. C. Bishop, T. C. Brown, and D. W. McCollum (1997): Using donation mechanisms to value non-use benefits from public goods, *Journal of Environmental Economics and Management*, **33**(2), 151–162.

Chilton, S. M. and W. G. Hutchinson (1999): Some further implications of incorporating the warm glow of giving into welfare measures, *Journal of Environmental Economics and Management*, **37**(2), 202–209.

Clark, D. E., V. Novotny, R. Grifin, D. Booth, A. Bartosova, M. C. Daun, and M. Hutchinson (2002): "Willingness to pay for flood and ecological risk reduction in an urban watershed", *Water Science and Technology*, **45**(9), 235–242.

Datascape & Communications Inc. (2004): Kurofune 2004 for Windows, Datascape & Communications Inc.

Daun, M. C. and D. Clark (2000): Flood Risk and Contingent Valuation Willingness to Pay Studies: A Methodological Review and Applied Analysis, Technical Report #6, Institute for Urban Environmental Risk Management, Marquette University, Milwaukee, WI.

Desvousgaes, W. H., F. R. Johnson, and H. S. Banzhaf (1998): Environmental Policy Analysis with Limited Information, Edward Elgar.

Freeman, A. M. III (1993): The Measurement of Environmental and Resource Value: Theory and Methods, Resources for the Future, Washington, D.C.

Greene, W. H. (2002): LIMDEP version 80, Reference Guide, Econometric Software, Inc.

Greene, W. H. (2003): Econometric analysis, Prentice-Hall International, Inc.

Hidano, N. and T. Kato (2000): A comparison of tax and donation formats in contingent valuation surveys: The case of Japan, Discussion Paper No. 00-1. http://www.soc.titech.ac.jp/library/discuss/text/dp00-1.pdf (in Japanese with English abstract). Accessed on May 19, 2005.

Krzanowski, W. J. I. and F. H. Marriott (1998): Multivariate Analysis: Classification, Covariance Structures & Repeated Measurements, Oxford University Press.

Luce, R. D. (1959): Individual Choice Behaviour: A Theoretical Analysis, New York, Wiley.

Mangione, T. W. (1995): Mail Surveys: Improving the Quality, SAGE publications.

Manski, C. (1977): The structure of Random Utility Models, Theory and Decision 8, 229–254.

Mcfadden, D. (1973): Conditional logit analysis of qualitative choice behaviour, Frontiers in Econometrics, P. Z. ed., New York, Academic Press.

Shabman, L. and K. Stephenson (1996): Searching for the Correct Benefit Estimate: Empirical Evidence for an Alternative Perspective, *Land Economics*, **72**(4), 433–49.

SPSS Inc. (1999a): SPSS 10.0J for Windows, SPSS Inc. (in Japanese).

SPSS Inc. (1999b): Amos 4.0 for Windows, SPSS Inc. (in Japanese).

Thunberg, E. M. (1988): "Willingness to Pay for Property and Nonproperty Flood Hazard Reduction Benefits: An Experiment Using Contingent Valuation Survey Method", Diss. Virginia Polytechnic Institute and State University.

Thurstone, L. (1927): "A law of comparative judgement", *Psychological review*, **4**, 273–286.

Wakui, Y. and S. Wakui (2003): Covariance Structural Analysis, Nippon Jitsugyo Publishing Co., Ltd. (in Japanese).

Zhai, G., T. Fukuzono, and S. Ikeda (2003): An Empirical Model of Efficiency Analysis on Flood Prevention Investment in Japan, *Review of Urban & Regional Development Studies*, **15**(3), 238–254.

Zhai, G., T. Fukuzono, and S. Ikeda (2005): Modeling Flood Damage: Case of Tokai Flood 2000, *Journal of the American Water Resources Association*, **41**(1), 77–92.

Zhai, G., T. Fukuzono, and S. Ikeda (2006a): An Empirical Model of Fatalities and Injuries Due to Floods in Japan, *Journal of the American Water Resources Association*, **42**(4), 863–875.

Zhai, G., T. Sato, T. Fukuzono, S. Ikeda, and K. Yoshida (2006b): Willingness to pay for flood risk reduction and its determinants in Japan, *Journal of the American Water Resources Association*, **42**(4), 927–940.

A better integrated management of disaster risks: Toward resilient society to emerging disaster risks in mega-cities, Eds., S. Ikeda, T. Fukuzono, and T. Sato, pp. 89–107.

New Mode of Risk Governance Enhanced by an e-community Platform

Toshinari Nagasaka

1 Introduction: Risk Governance, Local Community and Information Communication Technology (ICT)

Local communities face a wide range of underlying risks from natural disasters such as earthquakes, floods, and landslides. To reduce these, it is essential for these communities to prepare for them in advance and expend sufficient effort toward reducing damage in emergencies through the cooperation and mutual support of residents. However, in Japan social transformations, such as aging, urbanization, and changes to the employment structure, have made local-community social systems vulnerable to possible disaster risks by reducing cooperation between residents in neighborhoods. To tackle this problem, it is necessary for these residents to become active by enhancing means of participation where all residents and stakeholders jointly carry out properly informed choices concerning risks (National Research Council, 1996).

To deal with the uncertainty of disaster risk and to improve the safety and the relief of the local community, risk governance becomes important. The risk governance features the disaster prevention which utilized the social capital such as the mutual support among the residents, the social-network by NPO, the corporate citizen and so on which is enhanced by the local community at both normal and emergency times and the damage reduction by it (Fig. 1). It is different from the systematic measure which depends on regulatory institutions of preventing against disasters. To respond to newly emerging disaster risks that have nature of complexity, uncertainty and ambiguity, risk governance as a holistic and integrative strategy becomes indispensable.

Some local communities throughout Japan have suffered from large-scale natural disasters that have resulted in more damage than they had anticipated. The lessons obtained from these experiences are important for disaster preparedness that enhances mutual support by the local community and relief provided by the social network of volunteer citizens. In recent cases, many NPOs and volunteers have played an active part in the process of recovery from flood damage caused by the Niigata cloudburst, which occurred on 13 July, 2004, and the Niigata Chuetsu earthquake, which occurred on 23 Oc-

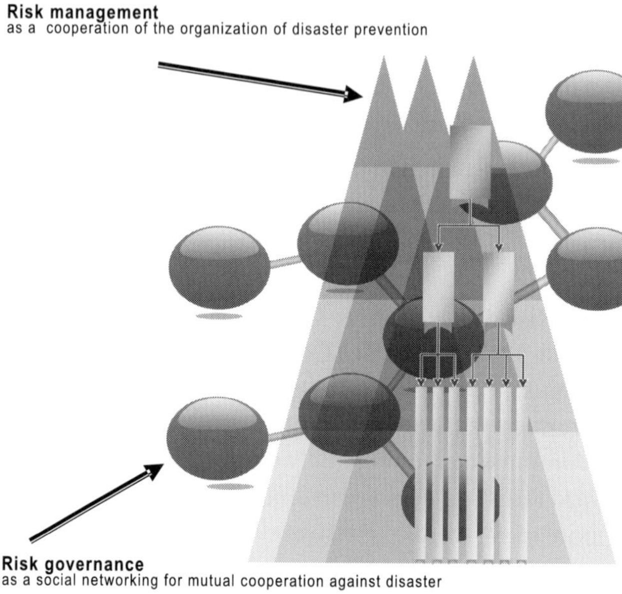

Fig. 1. Image of Risk governance.

tober, 2004. Some of the NPOs coordinating volunteers during the flooding used Web-log systems to send information such as relief needs from the affected area. After this earthquake, friends and relatives outside the affected area were able to check on the safety of residents using a commercial social networking system and they could offer encouragement through online text-based conversations. In this way, Web-log and social networking systems have played vital roles, enabling residents to cooperate during disasters and volunteers outside the affected areas to support victims.

In order to support the implementation of "risk governance", two types of platforms are at least necessary. One is the platform for providing the interoperability of disaster risk information scattered in different organizations. The other is the e-community platform for assisting formation of communities with the enriched social capital.

However, excluding our experimental efforts in Japan, there have been no practical approaches based on risk governance and very little has been done to implement it in the local community in terms of utilizing information communication technology. For this research project, we have been developing a prototype of an e-community platform as a place on the Internet to support the formation of communities, the creating their activities, and the implementation of risk governance through communication (Kaneyasu *et al.*, 2004; Shinkai *et al.*, 2002). This prototype is composed of a Web-log system and a geographic information system. Resident and citizen groups can use them synthetically with a general browser and a cell phone. We have been implementing an experiment using this prototype since December 2004 in co-operation with the local government and the residents of Shimada City which is located in the Shizuoka Prefecture of Japan.

Using the e-community platform, which is in daily use in the local community in this way, neighborhood associations and citizen groups in the area can also share the characteristics of disasters, previous disaster experiences by residents, hazard information, and risk information about the area. Administration information regarding disaster regulation or management measures, and associated scientific knowledge by experts can also be provided through the platform. Residents, NPOs, public administration, and experts, could share urgent disaster data and information through mutual cooperation with nearby residents in the case of disasters, and it would be the ideal way to provide relief via the social network on the platform. We called the concept of building such a social system, which increases the level of disaster prevention through risk communication, risk governance in the local community.

NIED is cooperating with several local governments, several local NPOs, and residents in specific areas to plan, design, and implement a prototype of the e-community platform. We are carrying out social experiments associated with this concept of risk governance. At present, they are being conducted in Shimada City in Shizuoka Prefecture (http://www.community-platform.jp/portal/), Fujisawa City in Kanagawa Prefecture (http://www.cityfujisawa.ne.jp/e-map/), and Tsukuba City in Ibaraki Prefecture (http://www.e298.jp/info/). Further studies are planned for the Tokai district. Below are two reports on the present situation with the experiments in Shimada and in Fujisawa Cities about risk governance using the e-community platform and on future research programs.

2 Approach of Implementing Risk Governance through Risk Communication by the e-Community Platform

Some of the most important factors in risk governance, which needs to be holistic and integrative, consist of regulations by public administration, risk perception by individuals or the community, community planning, and the culture of disaster prevention. The approach to risk governance through risk communication, which is promoted by the e-community platform, proceeds through 15 steps. Steps 1 to 7 involve capacity building in the local community. Specifically, the third step becomes important for finding active participants who will play a core role in the risk governance process. Activities involving various human resources and groups that used to have no interest in protecting against disasters in the local community are reflected in the platform. In other words, this phase is a preliminary step for implementing risk governance. Steps 8 to 15 become actual implementation phases for risk governance through risk communication between active participants and stakeholders (National Research Council, 1989; Renn, 1991). It should also be noted that these phases include the sharing of knowledge with other areas outside the target community and include the process of supporting the continuity of risk governance.

1. Forming e-community platforms in local areas (one platform for every town and village).
2. Forming communities on the platforms where the activities of each are activated.
3. Visualizing human resources and social capital available in the area.
4. Cooperating during activities by the community engaged in different missions.
5. Accumulating and utilizing social capital in the area for problem solving in the local community (Putnam *et al.*, 1993; Fountain, 2001).
6. Promoting partnerships between citizens and the local government.
7. Solving capability problems in the community as they arise and sharing experience and knowledge, or risk sense, for problem-solving beyond the area.
8. Forming alliances between institutions such as disaster prevention agencies, research institutes to protect against disasters, local government, and universities to provide scientific information on hazards and risk information online.
9. Enabling NPOs to be in charge of moderating risk communication between residents, neighborhood associations, and voluntary organizations for disaster prevention.
10. Continuing a series of these activities by NPOs.
11. Spreading these activities to other areas through the platform on the Internet and networking activities by NPOs.

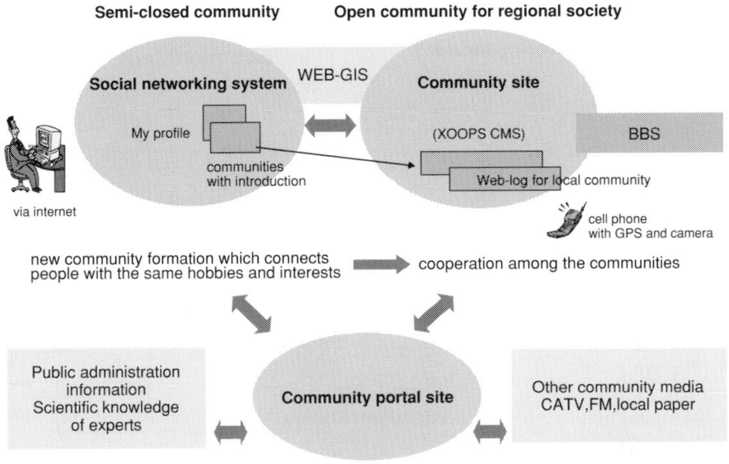

Fig. 2. Composition and functions of e-community platform.

12. Generating a wide range of regulatory or non-regulatory measures for risk reduction by the local community through risk communication.
13. Widely acknowledging the process of risk governance throughout the local community and advances in risk management by the local government.
14. Establishing a strategy of risk governance based on informed choice by the residents themselves and promoting cooperation of the local community.
15. Enabling consensus building between residents and stakeholders to reach at a certain level concerning the disaster risk and quality of social decisions.

3 Composition and Functions of the e-community Platform as a Prototype System

The prototype system is composed of a Web log system, a geographic information system, and other systems in relation to communication tools, as shown in Fig. 2. Resident and citizen groups can seamlessly access it from a common Web browser or cell phone. The prototype for Shimada City, a typical local city in the middle of Japan, was constructed by using an open-source content management system called XOOPS, which has a Web log function. The prototype for Fujisawa City, a suburban municipality adjacent to the

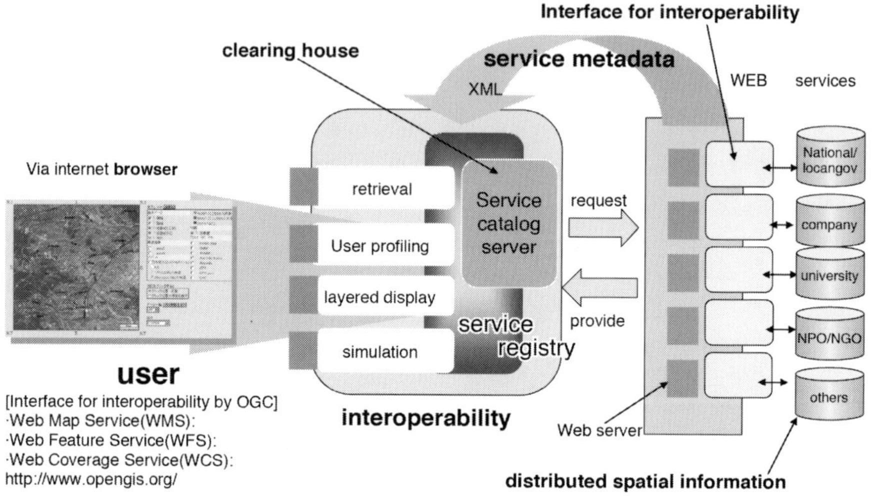

Fig. 3. Interoperability of spatial information.

Tokyo metropolitan area, was built using a self-developed content management system with the function of a geographic information system. Both prototype systems used by Shimada and Fujisawa cities were different according to whether or not they used XOOPS, but the functions that users could use were approximately the same. The composition of the prototype system using XOOPS is outlined below.

The system works in an environment constructed with open source software, such as Linux (operating system), PHP (protocol), MySQL (data base), and Apache (Web server). However, there is no open source module for displaying geographic information in XOOPS. The prototype system thus uses a commercial geographic information system (GIS), which supports interoperability (Fig. 3). It was incorporated into XOOPS as a plug-in, so users can access it by simply logging onto XOOPS. Its interoperability enables it to dynamically deliver the map data stored on all servers on the network based on user requests.

Users in neighborhood associations and citizen groups can easily set up a home page for a community without the need to know any computer languages, such as HTML, and can also prepare a server. They can quickly upload text and photographs to a community page from a PC or a cell phone. They can also upload text and photographs with spatial information added to the base map. Users can then find places related to the text and photographs on the map, or they can read text and view photographs related to places on the map.

XOOPS has a function that enables the administrator to limit the uploading of text and photographs and to add comments based on the access rights of users. All user groups can control the use of entries and comments with this function. Backtracking to guest users based on the management policy is also possible. The system can thus be managed as groupware, which only members can use.

The system can also be used to interact with the public by opening up access to anyone. XOOPS has modules for a bulletin board system (BBS), a photographic archive, a calendar, and a questionnaire. The open-source developer community developed these modules, and anyone can use them for free. We plan to extend the capabilities of some of these modules tailored to the e-community platform. We have already added a function for entering information from a GPS cell phone, and will add functions for printing poster-size maps and drawing course lines and area coverage during the coming year. Following that, we also plan to strengthen cooperation between the main system and the social networking system to facilitate mutual communication by concerned users (Fig. 4).

4 Implementation Experiments on the e-community Platform in Shimada City

4.1 Overview of Shimada City

Shimada City is approximately situated at the center of the Shizuoka Prefecture. The city has an area of 195.40 km^2. There are mountain ranges bounding the northern part of the city and large-scale tea plantations stretching along its southern flank. After having merged with neighboring Kanayacho in May 2005, the population of Shimada City grew to 98,591 with 32,106 households. About 10% of the working population is engaged in agriculture, 40% in manufacturing, and 50% in the service industry. One large river, the Ohigawa, and many tributaries flow through the city. The rich forest resources in the Ohigawa basin have led to the development of timber-related industries and paper is still one of the basic industries in the city.

4.2 Specific nature of disaster risks in Shimada City

There have been no catastrophic natural disasters in Shimada City in recent years. The Ohigawa river came under central governmental control in 1958. Repairs to its embankments and the seawall and an environmental improvement project on the floodplain in its lower reaches, have been carried out and, upstream, a dam is being built for flood control. The risk of flood-

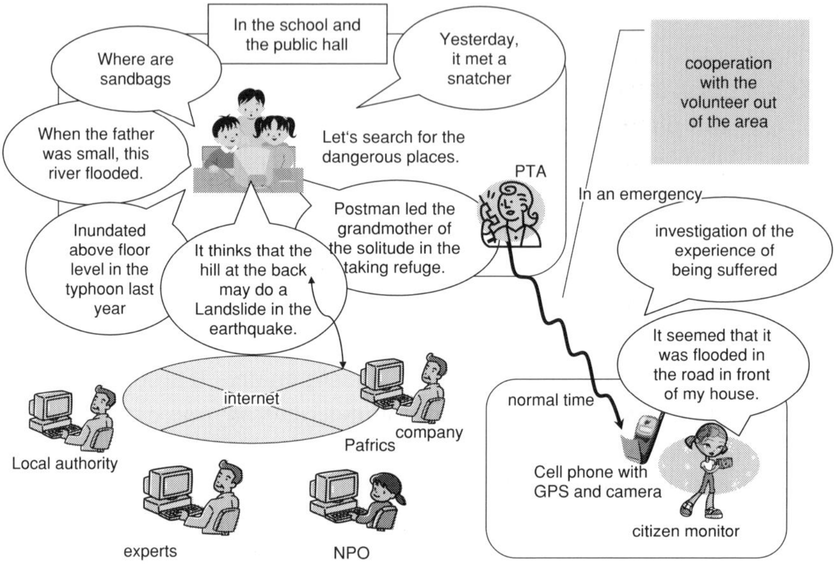

Fig. 4. Risk communication image in the local community.

ing caused by the Ohigawa has been reduced by these flood control measures. Historically, Ohigawa flooded every time it rained heavily and threatened the lives and fortunes of residents. To prepare for flood damage, residents constructed houses in the forms of ships and formed groups of these in triangles. Some of these ship-style houses still remain, but such a disaster-driven culture can hardly be seen these days.

A cloudburst, which was the biggest natural disaster in recent history, occurred in 1959. The daily rainfall reached as much as 376 mm. The banks of small and medium sized rivers in the city were destroyed. Numerous houses, rice fields, and other fields were flooded. Three people were injured and one went missing. There was a considerable amount of damage to buildings: 8 cases of complete collapse, 14 cases of partial destruction, 8 buildings washed away, 3,000 cases of inundation above floor level, and 6,000 cases of inundation below floor level. As far as earthquakes are concerned, Shimada City has not experienced significant damage in recent years. However, the city as part of the Shizuoka Prefecture has been included in the "countermeasures against earthquake disasters as a reinforcement area" for the expected Tokai earthquake by the Central Disaster Prevention Council of the national government. The Tokai earthquake is expected to be catastrophic and occur with an epicenter in the Suruga trough. In fact, large-scale earthquakes of magnitude eight have occurred about every 90–151 years in the Suruga trough in the Shizuoka Prefecture, but there has not been a disastrous earthquake in the trough since 1854. According to the government's seismic survey committee, the probability of a disastrous earthquake occurring in the next 30 years is estimated to be 84%.

4.3 Conditions for the experiment

Urbanization and aging in Shimada City in recent years have made such local communities as neighborhood associations vulnerable. The local government and residents have a vague sense of crisis about such changes in society. Although it varies within areas of the city, some are anxious about flood damage from small and medium-sized rivers, and one concern is the breakdown of social norms such as an increase in crime. Another concern is cooperation between newcomers and established residents. In April 2004, a research organization for empowering local communities with ICT was initiated by the local government to include those concerned about various types of social issues involving disaster risks. The organization was composed of university researchers, residents, and local companies. NIED participated as a member of this research organization and we cooperated with it by developing the specifications for the prototype e-community platform to empower local communities. After we implemented the prototype system in November 2004, we initiated a series of social experiments with it since 24 December, 2004. This experiment was scheduled to run for five years.

4.4 Progress to date

The first social experiment was started on 24 December, 2004, when the prototype system, called "e-community Shimada", was opened on the Web as a community site. We called citizen groups, residents, their friends, and students at colleges in the community to participate in e-community Shimada, which was designed to support local activities through the use of information technology (Fig. 5).

As of 7 September, 2005, 15 groups have been participating. The groups have various interests including nature, the environment, childcare, music, food, flood damage, the media, cultural assets, and area resources. Two of the groups represent the local community, i.e., nearby residents. One of these is called the "Ootsu area" group. Their pages on the e-community site cover a wide range of neighborhood association activities. For example, when heavy downpours occur, the pages become a tool for sharing vital information, such as the current water level of the river. All groups provide a page for guests to leave comments for exchanging information. During June 2005, about 30,000

Fig. 5. Sample of e-community Shimada.

pages were viewed with about 152,000 clicks. As of 13 July, 2005, 394 comments had been compiled.

The activities of all groups are introduced at the information-sharing portal to the e-community Shimada. There, users can quickly check the latest entries for all groups by using the mechanism for the rich site summary. There is also a moderated BBS at the portal to support communication between members of all groups. The moderator introduces the activities of all groups on bulletin board pages to promote mutual interaction among groups.

Meetings to promote group interaction and to encourage participation in the e-community Shimada are hosted at the activity base of each group in turn. These off-line meetings sometimes go beyond simply sharing mutual information to discussing problems associated with various groups. Moreover, these meetings are sometimes made more attractive by serving noodles and barbecued meat through cheap, voluntary systems. One researcher works in an office near Shimada Station that is referred to as an e-community café. It serves as an alternate base for not only his group but other groups as well (Fig. 6). The researcher frequently hosts workshops on system usage there.

Fig. 6. Risk communication workshop and offline meeting with users of e-community Shimada.

5 Discussion: Better-Risk Governance in Local Communities by means of e-community

One of the important purposes of this social experiments was to explore new methods of communicating information to ordinary residents who had little interest in possible disasters in their area, so that they could better understand the risks. Another was to support residents and NPOs so that they could design social networks to provide mutual help.

We are now proposing a new method of risk communication based on risk awareness and citizen participation that supports mutual communication between resident and citizen groups and NPOs while providing scientific knowledge and experience on disaster risk based on a year-round communication scheme. Several times a year, the local residents' group hosts a workshop on recognizing risks in the area. In addition to the risks of natural disasters, there are more common types of risk in terms of security, such as mugging and pick-pocketing. Residents at the workshop discuss possible measures for preventing these based on their experiences and/or hearing evidence. After this,

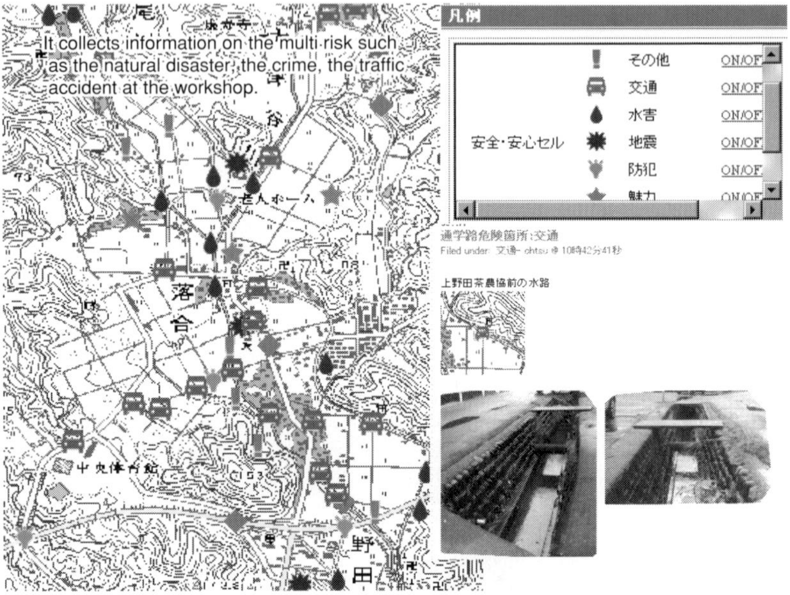

Fig. 7. Sample of hazard mapping by residents using e-community Shimada.

the residents collect information or evidence about hazards or risks posed by such security issues in the area (Fig. 7). The coordinator then gathers this information and posts it on the home pages of the e-community system using the GIS tools to make it available to the public. The local government, NPOs, and experts may add specialized information or comments related to the information collected by the residents who can request information and advice from the experts about disaster-prevention measures in the area. In this way, citizen groups, NPOs, or experts with specialized knowledge on disaster prevention may contribute to better governance of security issues. Moreover, they can add information or reliable evidence to the disaster experience database that NIED has developed and has made available on the Internet, and residents or NPOs can use it for their risk communication workshops.

In addition to improving risk communication between concerned groups and residents, all participating groups or individuals can report the present status of typhoons or cloudbursts, for example, in the area on their pages. This information could be used to help the elderly and other residents at risk to take shelter quickly. The information accumulated in the database is also organized for use in disaster education by the resident groups and NPOs. Moreover, resident groups can conduct emergency drills to ensure that information, such as

Fig. 8. Emergency drills using the e-community Shimada Aug 2005.

the status of disaster relief and the demand for various types of help, can be smoothly entered into the e-community Shimada in the form of photographs and text from their cell phones. Incidentally, residents may utilize this function for informing one another about local events such as festivals, assemblies, and meetings and thus become more accustomed to using it (Fig. 8).

Elementary and junior-high-school students can also make use of the e-community Shimada as a tool for understanding the local community as part of their school programs. They can learn which particular areas may be of risk and about shelters or spots where disaster-prevention facilities are located or people are stationed. Residents can interact with people outside the area using the e-community Shimada to promote cooperation during times of disaster. People who are willing to provide assistance can describe their capabilities in the social networking system, enabling neighborhood associations and NPOs to be identified and people to be contacted who can provide necessary assistance.

These processes comprise a model for risk governance in local communities through mutual communication based on the e-community Shimada.

Some functions of the model have already been implemented, including residents themselves collecting and disseminating risk information and conducting emergency drills.

6 Implementation experiments in Fujisawa City

6.1 Overview of Fujisawa City

Fujisawa City is located in the south of the Kanagawa Prefecture and is about 51 km from central Tokyo. It faces Sagami Bay on the Pacific Ocean and has an area of 69.51 km^2 extending 12 km north to south and 6.5 km east to west. There is an elevation of 40–50 m above sea level in the north of Fujisawa. In the south, lie alluvial lowlands. The Sakai and Hikichi rivers flow from the north of the city into Sagami Bay. Fujisawa is a popular metropolitan area and has a population of approximately 390,000.

The city is a typical suburban area where industrialization started in the early 1960s. It has developed since then to have a solid social infrastructure. However, as many of its production facilities have moved overseas, it is becoming a center for R&D.

6.2 Specific disaster risks in Fujisawa City

The most dreaded risks that Fujisawa City faces are the so-called Tokai earthquake, earthquakes in the west of Kanagawa Prefecture, earthquakes in the South-Kanto plain, and inland earthquakes with vertical thrust. Tsunamis caused by earthquakes have been guarded against in coastal areas. The north is exposed to the risk of flooding by typhoons or local heavy downpours, and mudslides down steep slopes. When Fujisawa City was devastated by Typhoon No. 22, 2004, 311 families were flooded out because of drains blocked by mudslides.

6.3 Trends of citizen participation via Internet in Fujisawa

Fujisawa City introduced an electronic bulletin board system (BBS) in 1997 to facilitate citizen participation and to create a new type of community. It is managed by a committee, which is composed of citizens who were recruited. This BBS enables residents to propose administration policies that are formed through an exchange of views directly with the mayor. This system is also used by residents to share information about areas of concern. Nationwide, it is a remarkable cutting-edge system and typical of e-democracy in Japan. About 3,000 citizens subscribe and are currently using it to facilitate the participation of citizens in city administration (Kaneko, 2004).

Fujisawa City has another community ware for citizens called Den-en map, which has been making progress since 2003. This system has a Web-log assisted by a Web-GIS function. All citizens can easily create their own maps using this system and make them public on the Internet. Because it has the Web-log feature, guests can post comments or replies to other residents in addition to browsing maps. Residents can post messages to Web-GIS maps with photographs using a GPS cell phone with a camera. A city-planning map and aerial photo are provided as background for posting. This system is managed by a promotion committee, which is composed of citizens, and university, industry, and city administration employees. Some citizen groups are regularly investigating the barrier-free status of public facilities and commercial establishments. They have created a barrier-free map from the research results obtained by using this system and are making it available on the Web-GIS.

Other maps have included gourmet mapping by citizen groups, mapping of neighborhood associations, crime prevention mapping, and nature-view mapping. As of March 2006, the number of subscribers was 87 and the number of maps was 34. The number of accesses in January 2006 was about 52,000 and about 43,000 in February.

The Web-GIS is equipped with a global standard interface for interoperable map content through the Web Mapping Service. Through this interface, residents can access hazard maps and engage in risk mapping, which NIED is scheduled to provide dynamically in real-time.

6.4 Implementation experiment in Fujisawa City

NIED and Fujisawa City have been promoting joint research on the risk management of natural disasters since 2004. NIED started to deliver rainfall information obtained by high-precision radar in real-time with the Den-en map system in 2005. The level of inundation as a result of flooding can be simulated using this radar rainfall in terms of a spatial resolution of 10 m. The data are being delivered on a trial basis with the Den-en map system that citizens are using. NIED did a field survey to evaluate simulation with the results obtained by coordinators of a disaster relief volunteer group and residents who were interviewed in a survey on damage done by a past typhoon (Fig. 9).

There is a group of schoolchildren who are walking to observe the city. They have confirmed the dangerous places and the local evacuation spaces during their walks in the city. They subscribed information on the dangerous parts onto the electron mapping system by using GPS-compatible cell phones. At the workshop which was implemented after the walking, some children could not confirm the condominium which is specified by the evacuation space of the tsunami. Also, other children stated the opinion that the condominium cannot be used for the evacuation space for the auto lock. The technique of such risk communication will be useful for improving the consciousness of the

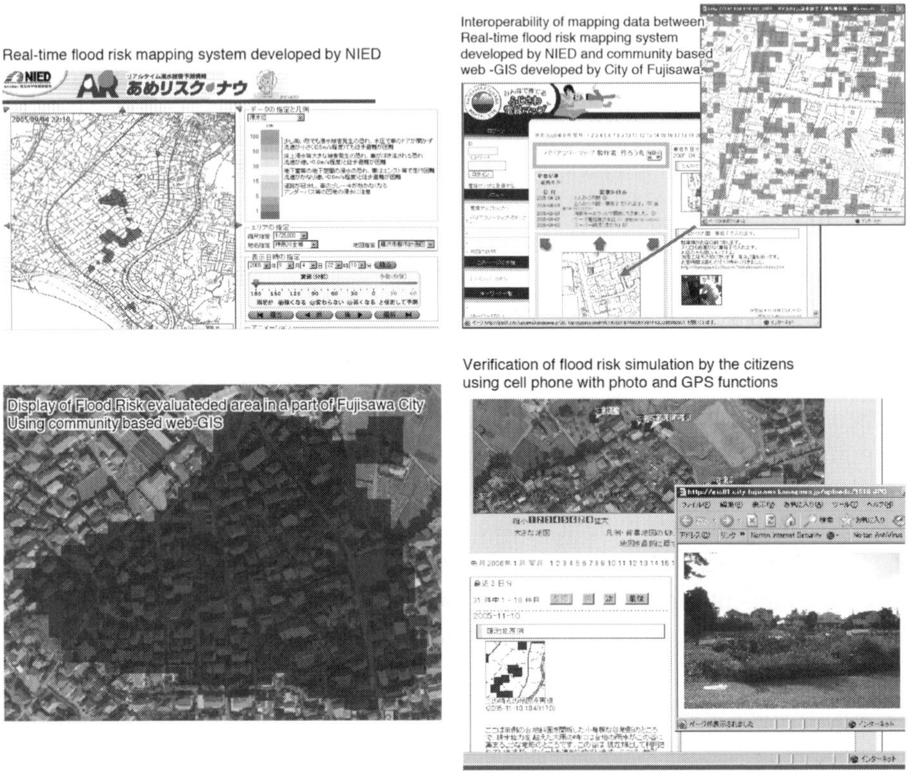

Fig. 9. Field survey to evaluate simulation of flood risk by disaster relief volunteer group using e-community.

resident of the protection against disasters and will be useful for evaluating in the effectivity of the contingency planning of the present situation (Fig. 10).

6.5 Progress to date

NIED is recording the results from a field survey and resident interviews with documents and photographs using a Web-log system that has a Web-GIS function and is preparing to make them public on the Internet. This demonstrates the possibility of residents themselves greatly enhancing risk communication by means of hazard maps that NIED provides to the local community through the e-community platform. NPOs have become important in implementing risk governance in the local community in this way. NPOs in cooperation with experts, local governments, and residents, play the role of facilitators in the process of risk communication when there are no emergencies. Neither volunteer organizations nor residents who can become victims

Fieldwork and workshop by school children who utilize cellular phone
With photo and GPS functions and community based WEB-GIS No.1

Fieldwork and workshop by school children who utilize cellular phone
With photo and GPS functions and community based WEB-GIS No.2

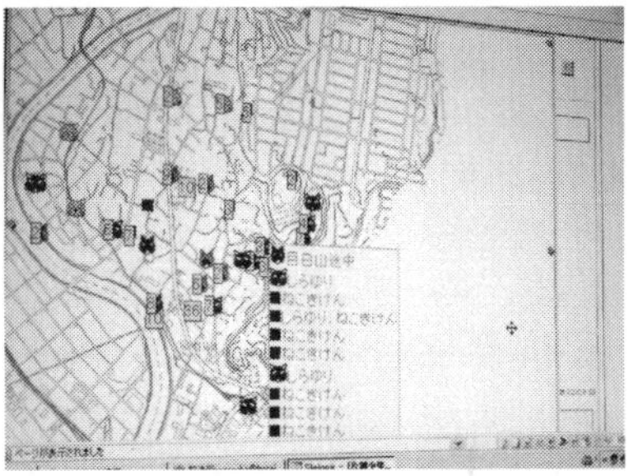

Fieldwork and workshop by school children who utilize cellular phone
With photo and GPS functions and community based WEB-GIS No.3

Fieldwork and workshop by school children who utilize cellular phone
With photo and GPS functions and community based WEB-GIS No.4

Fig. 10. Fieldwork and workshop by school children using e-community.

themselves can be expected to assume roles as facilitators or moderators in disaster-risk communication. Through workshops or meetings, in addition to communication through the network, local residents can share experiences they have suffered in the past, understand possible causes of disasters, and review the variety of measures available to reduce risk in the future. Also, residents can discuss what level of disaster risk they can tolerate or accept based on shared information and experiences.

As previously mentioned, part of the mechanism for sharing information that NPOs, residents, and local governments in concerned areas can utilize during catastrophes, and the risk information that NIED provides through the e-community platform have already been implemented.

7 Conclusion

The effectiveness of this model of risk governance in the local community could be evaluated through participant observations of the activities by citizen groups, quantitative analysis of the use-logs of the e-community, qualitative analysis of comments and photographs uploaded to the e-community, interviews with the participants, and questionnaires sent to residents.

The elements evaluated in terms of the effectiveness of risk governance were:

1. groups in the network community,
2. the number of citizens participating in the e-community platform,
3. concerns about hazards and risks, besides those of natural disasters,
4. the number of interactions between resident and citizen groups and NPOs or experts with specialized knowledge on disaster prevention measures, and
5. the number of cooperative activities in the local community that increased the scope of the social network.

Both social experiments in Shimada and Fujisawa cities were firstly conducted to find out residents' perceptions of the risk of flood damage, and were then extended to the following:

1. In addition to flood damage, the model dealt with earthquakes, landslides, and tsunamis to enhance communication about risk in the local community.
2. In the experiment in Shimada, risk communication was primarily developed around residents, and in part, between them and the NPOs. In future, we intend to do research that promotes mutual communication about risk between residents and the local government, and other stakeholders such as industries.

3. To utilize the technology of Web-GIS and RSS with updated risk information, we plan to continue further experiments on the interoperability of the e-community platform in which participants utilize not only hazard information that the administration sends, but also risk-related information that residents send for.

References

Fountain, J. E. (2001): *Building the Virtual State, Information Change and Institutional Change*, pp. 64–82, 193–206, The Brookings Institution.

Ikeda, S. (2004): Risk Society and Planning Approach toward Social Governance on Technical and Environmental Risks, *Planning Administration*, **27**(3), 3–11 (in Japanese).

International Risk Governance Council (2005): White Paper on Risk Governance, http://www.irgc.org/irgc/projects/risk_characterisation/_b/contentFiles/IRGC_WP_No_1_Risk_Governance_(reprinted_version).pdf

Kaneko, I. (2004): *Challenge for e-democracy*, pp. 182–190, Iwanami Pub. (in Japanese).

Kaneyasu, I., Nagasaka, T., and Shinkai, I. (eds.) (2004): *The Guideline in Electronic Conference for Citizen's Participation*, pp. 138–159, Gakuyo Shobo (in Japanese).

National Research Council (1989): *Improving Risk Communication*, National Academy Press.

National Research Council (1996): *Understanding Risk: Informing Decisions in a Democratic Society*, National Academy Press.

Putnam, R. D. *et al.* (1993): *Making Democracy work*, pp. 163–185 (Social Capital defined in p. 167), Princeton University Press.

Renn, O. (1991): *Strategies of Risk Communication: Observations from Two Participatory Experiments*, pp. 457–481, in R. E. Kasperson and P. J. M. Stalle (eds.), Communicating Risks to the Public, Kluwer Academic Publishers.

Shinkai, I. *et al.* (2002): *Risk Communication Support System*, pp. 193–200, e-democracy as a regional strategy, Shogakkan (in Japanese).

Part II: Interdisciplinary studies of flood risk

A better integrated management of disaster risks: Toward resilient society to emerging disaster risks in mega-cities, Eds., S. Ikeda, T. Fukuzono, and T. Sato, pp. 109–119.

Uncertainty in Flood Risks and Public Understanding of Probable Rainfall

Shinya Shimokawa and Yukiko Takeuchi

1 Introduction

Flood risk data presented in hazard maps and other materials may be generally considered reliable. Such data, however, are not completely reliable, although they are obtained through calculations using methods and according to standards that are reliable to a certain extent. There also exists a gap in the understanding of data between professionals and the public. That gap may exacerbate the damage caused by floods. The Pafrics, which we developed in our project (Chapter 13), will be a useful tool for bridging this gap. In this chapter, we discuss the uncertainties involved in the determination of flood risks and the concept of risk assessment considering such uncertainties. Our present level of knowledge cannot easily accommodate such uncertainties. In relation to hydrological and hydraulic factors, numerical models are employed for such uncertainties (stochastic and/or statistical models, e.g. Bolgov *et al.*, 1998), but they are not yet perfect. We explain the existence of such uncertainties specifically using hydrological statistics as an example, which are used in flood prevention planning. The unpredictability in hydrological statistics, which is due to uncertainty and variability (Vose, 2004), and the difference between estimated probabilities of rainfall and earthquakes are also discussed. In addition, we present the results of questionnaire surveys regarding public understanding of probable rainfall, which is used in hazard maps.

2 Physical Processes of Flooding

Flooding is generally composed of four processes: (i) rainfall as input, (ii) runoff, (iii) flooding, and (iv) depth of inundation as output (Fig. 1). Each process is described below.

(i) Rainfall as input

Rain falls for certain reasons. (In grasping flood phenomena, rainfall is handled as an outright external force. The intensity of rainfall is estimated based on hydrological statistics using past rainfall data.)

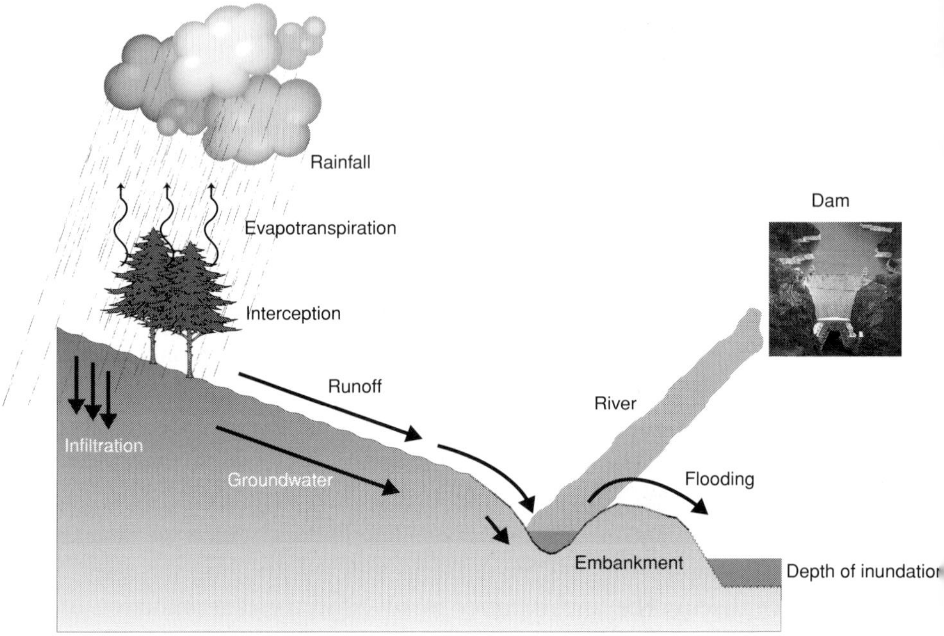

Fig. 1. Physical processes related to flooding.

(ii) Runoff

Not all of the rainfall flows off the ground surface. Some rainfall is intercepted by trees or infiltrates the ground and only some rainfall flows off the ground surface. The water infiltrating the ground also flows out to the surface some time later. Thus, rainfall does not match runoff at a given point in time. Identifying the percentage of rainfall that flows off the surface according to the characteristics of the ground surface or soil is therefore necessary.

(iii) Flooding

Overland water flows into rivers or onto roads. The routes of runoff are determined by elevation and obstacles such as buildings. Where rivers overflow their banks depends on the flow in the river, the shape of the river, and the strength of embankments. It is therefore necessary to identify where the runoff water concentrates and where rivers are most likely to overflow their banks.

(iv) Depth of inundation as output

The water overflowing banks flows into another area or is discharged by area drainage systems into sewers. Finally, the distributions of water level and changes in water level with time at the place of overflow and in surrounding areas are obtained. The depth of inundation determines the magnitude of flood damage.

3 Key Flood Risk Factors

Key determinants of flood risks are listed below (National Research Council, 2000).

- Hydrological factors
- Hydraulic factors
- Structural and geotechnical factors
- Material and construction factors
- Seismological factors
- Other geophysical factors
- Operation and maintenance factors

Hydrological factors include rainfall, flooding, and basin and channel data. Hydraulic factors are the characteristics of floodwater propagation and the equations and methods used to simulate such propagation. Structural and geotechnical factors refer to the geographical and geological characteristics of ground and soils. Material and construction factors mean the materials and methods used for constructing structures such as dams and embankments. Seismological factors include the destruction of dams, embankments, and other structures, as well as soil liquefaction due to earthquakes. Other geophysical factors are the behavior of ice in dam reservoirs or rivers, lightning, and tornadoes. Operation and maintenance factors refer to the operation of dams in emergencies and the maintenance of river systems.

4 Propagation of Uncertainties and the Present Response

Flood risk factors are interrelated. Uncertainties in a single factor may lead to greater uncertainties as they propagate even where other factors are perfectly modeled. Rainfall, runoff, depth of inundation, and the magnitude of damage are calculated in this order as parameters related to flooding. Small margins of errors in rainfall increase gradually as margins of error are accumulated in respective steps. The present level of knowledge has difficulty in handling such uncertainties perfectly. In relation to hydrological and hydraulic factors, models are being employed that provide for such uncertainties (stochastic and/or statistical model, e.g. Bolgov *et al.*, 1998), but they are not yet perfect.

One of the serious problems in disaster prevention planning is the lack of public understanding about the numerous uncertainties involved in flood risk estimation. The next section specifically explains how such uncertainties come about using, as an example, hydrological statistics which provide a basis for flood prevention planning.

5 Uncertainties in Hydrological Statistics

Hydrological statistics are used to estimate the intensity of potential rainfall based on past rainfall data. For example, the intensity of 100-year probable rainfall is obtained based on past rainfall data. This serves as a basis for developing flood prevention plans. In Class-A rivers, it is required that embankments can be strengthened to endure the 100- or 200-year probable rainfall expected in the area in question.

Hydrological statistics involve the following uncertainties.

(i) Uncertainty of probability
(ii) Uncertainty of data
(iii) Uncertainty owing to the use of a finite number of data
(iv) Uncertainty attributable to variation in probability estimation methods
(v) Uncertainty attributable to climate changes

Detailed explanations follow.

(i) Uncertainty of probability

There can be no assurance that 100-year probable rainfall events will occur only once in 100 years. The occurrence of such an event in any given year cannot ensure there is no 100-year probable rainfall event in the next year. Such events could occur in two consecutive years. Probability therefore involves uncertainty. There may, however, be misunderstanding among the public about this point in numerous cases. (This matter will be discussed again in following sections.)

(ii) Uncertainty of data

Data that are used to estimate probabilities always involve margins of error. The older the data is, the greater the margin of error. The probability of rainfall estimated based on error-prone data therefore involves uncertainties.

(iii) Uncertainty owing to the use of a finite number of data

Only a limited amount of data on past rainfall events is available. Those on rainfall events more than 100 years ago are particularly scarce. Probable rainfall estimated based on a limited amount of data involves uncertainties.

Table 1. Estimated probability of the Tokai heavy rainfall.

Case A	Kawata (2002)	At least once in 200 years
Case B	Mizutani (2002)	At least once in 1000 years
Case C	Ushiyama and Takara (2002)	Once in 40 137 years

(iv) Uncertainty attributable to variation in probability estimation methods

Probabilities are estimated by various methods. Logarithmic normal distributions are suitable for expressing annual extreme values over a relatively long term (a month or a season). Extreme-value distributions can properly represent annual extreme values over a relatively short term (a day or an hour). Non-annual data can be handled well using exponential distributions. The selection of the method, however, is basically arbitrary. No choice is regarded as mathematically wrong. Actually, though, results can vary greatly according to the method selected. The fluctuation is outstanding in areas of poor reliability; e.g., low-probability events such as 100-year probable rainfall events. Estimation of probable rainfall therefore involves uncertainties.

(v) Uncertainty attributable to climate changes

When estimating the amount of rainfall for a certain probability based on past rainfall amounts, it is assumed that climate remains unchanged from the past. Climate, however, does change, which in turn causes the mode of rainfall to change. Phenomena due to natural climate changes, such as cold summers and warm winters, are well known. Man-made climate changes, including CO2-induced global warming, have also become an issue. These uncertainties are inherent in the estimation of the probability of rainfall.

Finally in this section, let us consider an example of the calculation of probable rainfall. The estimated probability of the Tokai heavy rainfall that occurred on September 11, 2000 (daily rainfall: 428 mm, peak hourly rainfall: 97 mm) vary from once in approximately 200 years to once in 40,000 years according to researchers (Table 1). The variations are ascribable mainly to the duration of data collection (data volume) and the method of probability estimation. These variations cause an uncertainty of design for disaster prevention structures (e.g. embankments and dams) and a social problem of environmental destroy by the structures (Ohkuma, 2004).

6 Peculiarities and Unpredictability of the Probability of Rainfall

The probability of rainfall has different aspects from ordinary probabilities (e.g., those when rolling dice). Each face of a die is likely to come up once in six rolls, as has been true in the past and will be in the future. That is not always the case with rainfall. Suppose the probability of a certain amount of rainfall is 1/100 in a given year. If a certain amount of rainfall exceeding the expected amount occurs several times in the next 10 years, the probability of that amount of rainfall will no longer be considered to be 1/100; it will be higher. This is related to the uncertainties due to climate changes. Once the mode of rainfall changes, there will be no assurance that the probability of a certain amount of rainfall occurring in the future can be estimated based on the past rainfall amounts. Even without climate changes, calculating the probability of a certain amount of rainfall in a given year from the past rainfall amounts is not guaranteed to be accurate; it is simply based on an assumption. The issue is how to make such an assumption. This is closely related to the prior probability in Bayes' theorem in statistics. (For Bayes' theorem, refer to the Statistics Section, Department of Social Sciences, College of Arts and Sciences, University of Tokyo (1992).) Regarding the application of Bayes' theorem to stochastic prediction, Katayama (1975) and Matsumura (2004) point out some interesting considerations concerning seismic prediction.

Unpredictability is attributable to uncertainty and variability (Vose, 2003). Uncertainty means a lack of knowledge about phenomena. Variability refers to the accidental or stochastic action of phenomena. These two factors generally combine to create unpredictability. In the case of probable rainfall, uncertainty of data, uncertainty owing to the use of a finite number of data, and uncertainty due to variation of probability estimation methods are "uncertainty". Uncertainty of probability and uncertainty due to climate changes should be regarded as "variability". The unpredictability of climate changes may, however, be "uncertainty" ascribable to our insufficient knowledge. The probability of rainfall differs from that when rolling a die in these respects.

7 Difference between Probabilities of Rainfall and Earthquakes

The probabilities of earthquakes and rainfall events are based on the same concept. They, however, vary in three respects.

(i) To estimate earthquake probabilities, the Brownian passage time (BPT) distribution based on Brownian movement is frequently used. This is because earthquakes and Brownian movement share a physical property that can be described by a Markov process which depends only on the immediately preceding phenomenon. The BPT distribution, unlike other mathematical distributions, facilitates physical interpretation and exhibits an upward-sloping line, even over an infinite period of time. Rainfall is basically a Markov process independent of the past (like rolls of a die), so using the BPT distribution to estimate the probability of rainfall is not physically justifiable.

(ii) Earthquake probabilities are calculated only for selected earthquakes of intensities that are unique to the area. For example, in the southern Kanto area, magnitude-8 earthquakes occur more frequently than magnitude-7 earthquakes. The magnitude-8 earthquakes in the area contradict the power law that smaller earthquakes occur more frequently than greater earthquakes (a drop of one in the magnitude means an earthquake is likely to occur 10 times more frequently), and are regarded as earthquakes of an intensity unique to the area. The accuracy of probability is therefore increased by selecting for calculation only magnitude-8 earthquakes (which occur at an interval of approximately 200 years). This depends on the characteristic of earthquakes. That is, the intensities of past earthquakes up to approximately 600 years ago (approximately 2000 years ago in China) can be identified based on archival data on seismic motions or the destruction of buildings. Rainfall-induced runoff, on the other hand, depends on the ground surface condition at a given place in a given age, so estimating rainfall from the descriptions of rainfall or runoff in ancient documents is difficult. In addition, great earthquakes do not occur every year, so probabilities are calculated only for great earthquakes, but heavy rains occur every year, so the probabilities of rainfall are calculated every year based on the maximum annual rainfall. More accurate probabilities, however, may be calculated based on 100 leading rainfall events during the past 100 years rather than on the 100 maximum heavy rainfalls during the past 100 years.

(iii) For earthquakes, cumulative probabilities (e.g., the probabilities of magnitude-8 earthquakes occurring in 10 years, or in 30 years) are calculated based on the assumption of an underlying physical process (a process of releasing stresses and then accumulating stresses again). This is possible because earthquakes can be described using a Markov process that depends only on the state of an immediately preceding earthquake. Rainfall, on the other hand, is a Markov process independent of the past (like rolls of a die), so the idea of cumulative probability based on an assumed physical process is difficult to apply. The probability of a 100-year probable rainfall occurring in 10 years, of course, can be mathematically calculated, but no physical process is assumed. Even without any heavy rains in the next year, the probability of 100-year probable rainfall events occurring in 10 years remains the same as in the previous year (as in the case of rolls of a die). On the other hand, histories of water content in soils or groundwater volume are available, so an approach similar to that for earthquakes may be applicable to the estimation of landslide probabilities or groundwater-runoff volumes. In addition, in rainfall events, unlike earthquakes, direct observation of the source of the event (such as atmospheric conditions) is possible. For example, the probability of a typhoon causing heavy rainfall of more than x mm can be estimated based on the moisture contained in the air.

8 Flood Risks and Their Uncertainty

One of the greatest problems involved in the uncertainties of flood risks is that various flood forecasts are considered to be certain although they actually contain stochastic factors. One-hundred-year probable rainfall, for example, does not necessarily occur only once in 100 years[1]. It may occur several times in a short period of time. The occurrence of a flood therefore never guarantees relief for a certain time period. Daily preparedness therefore becomes important.

From a stochastic viewpoint, a general concept of flood risk can be expressed quantitatively as

$$\text{Risk} = (\text{Probability of the hazard}) \times (\text{Damage caused by the hazard}).$$

Note that the risk expressed above is neither a vague mood nor an abstract word, but can be expressed numerically using the probability of and damage caused by a hazard, and that knowing the scale of a risk and taking appropriate control measures help reduce the potential damage. This probability and the scale of a risk, however, contain numerous uncertainties.

Particularly when estimating the magnitude of damage, the scale and definition of damage, and the value judgment or psychological factors of the organizations or individuals evaluating risk should be taken into consideration. Numerous uncertainties are also involved in risk evaluation by organizations or individuals. It may be natural for local residents who have suffered from large-scale flood damage over a long time to demand that the government improve rivers and construct estuary barrages or seawalls. There have, however, been vigorous campaigns calling for the protection of the natural environment, leading to protests against river improvements on numerous occasions. The former group focuses on the risk of flooding while the latter group focuses on the risk of degrading the environment. Differences in viewpoints, interests, or values cause such differences in focus. In addition, even among those paying attention to the same risk, the scale of acceptable risk may vary according to individual viewpoints.

[1]For example, the probability of a 100-year probable rainfall occurring at least once in the next 30 years is relatively high at approximately 26% ($1 - (99/100)^{30} = 0.26$), although this is contrary to common knowledge.

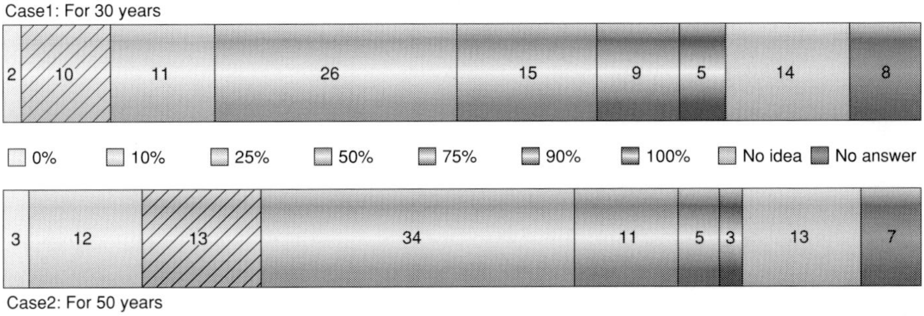

Fig. 2. What do you think is the probability of a 200-year probable rainfall occurring once in 30 or 50 years? (Nagoya City).

9 Questionnaire Surveys to Assess the Recognition of Uncertainty and the Understanding of Probable Rainfall

Questionnaires were distributed to residents to assess their understanding of probable rainfall, which is used to prepare flood hazard maps.

Questionnaires were distributed in Nagoya City and Nishibiwashima Town which suffered severe flood damage on September 11, 2000. Both of the local governments developed a flood hazard map after the flood and distributed copies to residents in 2002. Questionnaires were distributed to 3000 households in the areas expected to be inundated in the flood hazard map. Responses were obtained from 644 households.

The questionnaire included two questions about probable rainfall that residents had to understand to grasp the zones vulnerable to inundation specified in the hazard map. The questions were "What do you think is the probability of a 200-year probable rainfall occurring in the next 30 and 50 years?" (The right answers were 14 and 22%, respectively.) In Nagoya City, 10% of respondents provided the right probability for the next 30 years, and 13% gave the right answer for the next 50 years (Fig. 2). The corresponding percentages were 5 and 15% in Nishi-biwashima Town (Fig. 3).

The results indicated that residents had insufficient knowledge about the concept of probable rainfall defined by the expert although this knowledge is essential in understanding which zones are vulnerable to inundation, as shown in the hazard map.

10 Closing Remarks

Numerous uncertainties are involved in the determination of flood risks. In flood risk assessment, differences in viewpoints, interests, or values re-

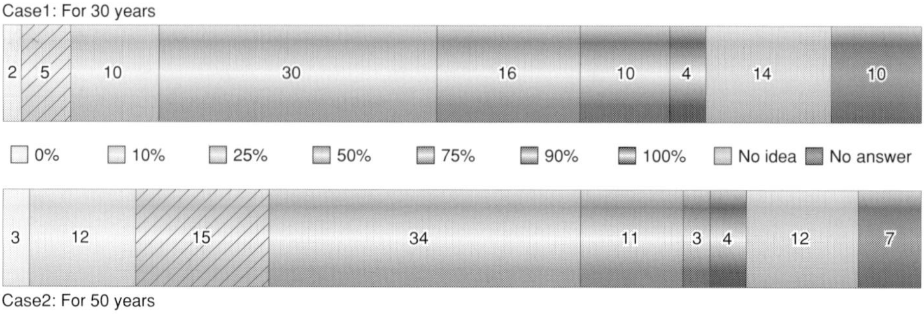

Fig. 3. What do you think is the probability of a 200-year probable rainfall occurring once in 30 or 50 years? (Nishi-biwashima Town).

sult in differences in risk focus. Each individual should determine what he or she considers an acceptable risk and take appropriate action while keeping the above in mind. The questionnaire survey results presented in Section 9 show that most residents have a poor understanding of probable rainfall. Proper knowledge about probability is indispensable to understand flood disaster risk, especially, understanding of the content on uncertinties involved in the mid/long-term probability. Support should be provided through workshops to help the public properly understand the uncertainties inherent in determining flood risks.

Acknowledgments. Regarding the discussions in Section 7, a lecture given by Dr. Masajiro Imoto (Principal Senior Researcher, National Research Institute for Earth Science and Disaster Prevention (NIED)) at NIED and subsequent discussions with him, were highly thought-provoking. We would like to express our gratitude to Dr. Imoto.

References

Bolgov, L. M., L. Gottschalk, I. Krasovskaia, and R. J. Moore (eds.) (1998): *Hydrological Models for Environmental Management*, Kluwer Academic Publishers, Dordrecht.

Katayama, T. (1975): Stochastic approach to seismic activities and risks, *Journal of Institute of Industrial Science*, **27**, 185–195 (in Japanese).

Kawata, Y. (2002): 2000 Tokai heavy rainfall and its damage, *Journal of Natural Disaster Science*, **21**, 8–14 (in Japanese).

Matsumura, S. (2001): Problems in evaluating earthquake occurrence rates based on a few examples, *Jishin (Journal of the Seismological Society of Japan)*, **54**, 45–46 (in Japanese).

Mizutani, T. (2002): *Natural Disasters and Science of Disaster Prevention*, University of Tokyo Press (in Japanese).

National Research Council (2000): *Risk Analysis and Uncertainty in Flood Damage Reduction Studies*, National Academic Press, Washington, D.C.

Ohkuma, T. (2004): "Basic High Water" prevents "No dam"—wandering Japanese river improvement plan, *Sekai*, **10**, 123–131 (in Japanese).

Statistics Section, Department of Social Sciences, College of Arts and Sciences, University of Tokyo (1992): *Statistics of Natural Science*, University of Tokyo Press (in Japanese).

Takeuchi, Y., K. Takao, S. Shimokawa, T. Satoh, T. Fukuzono, and S. Ikeda (2005): Verification of flood literacy learning support tools, Report of National Research Institute for Earth Science and Disaster Prevention, No. 67, 63–71 (in Japanese).

Ushiyama, M. and K. Takara (2002): Characteristics of 2000 Tokai heavy rainfall based on heavy rainfall records since 1901, *Journal of Natural Disaster Science*, **21**, 145–159 (in Japanese).

Vose, D. (2000): *Risk Analysis: A Quantitative Guide*, 2nd Edition, John Wiley & Sons, England.

A better integrated management of disaster risks: Toward resilient society to emerging disaster risks in mega-cities, Eds., S. Ikeda, T. Fukuzono, and T. Sato, pp. 121–134.

Public Perception of Flood Risk and Community-based Disaster Preparedness

Tadahiro Motoyoshi

1 Introduction

To carry out participatory flood risk management in local communities, it is necessary to understand how residents perceive flood risks in their areas and what type of consciousness residents have regarding disaster preparedness actions. Since not all people are so tolerant of flood risk in the areas where they live, it is important to identify which factors affect the degree of acceptance of people to flood risk. In Japan, community organizations such as volunteer fire organizations and flood fighting organizations play a role in preventing disasters. It has been pointed out, however, that the ability of communities to prevent disasters has declined as nuclear families increased, traditional communities declined, and solitary, live-alone old people increased. In order to perform early relief activities quickly in the event of a disaster, it is necessary to make local disaster prevention efforts on a daily basis. The public as well as the administrative authorities also need to participate and cooperate to protect their communities so that their efforts should be reflected in administrative planning. In order to find ways to involve local residents in disaster-prevention activities, there is a need of conducting a study to identify factors contributing to the participation of local residents in disaster prevention activities. This chapter examines consciousness in accepting flood risks and the intention to participate in community-based disaster preparedness activities and carries out studies on each causal model based on questionnaire surveys.

2 Public Acceptance of Flood Risks

2.1 Factors to determine acceptance of flood risks

One of the characteristics of recent flood disasters in Japan is that there is an extremely huge economic loss although the number of human casualties is declining. It is frequently pointed out that one reason for people's lack of preparedness against flood disasters is that there is an inappropriate perception about flood risks.

It has been shown in most studies on natural disaster that it is difficult for people to appropriately perceive natural disaster risks (Slovic *et al.*, 1974). For example, people tend to perceive flood disasters as periodic phenomena instead of as probable and random phenomena. Furthermore, people tend to believe that if a major flood disaster occurs in a certain year, no major flood disasters will occur for some time after. In addition, many people believe that when levees, dams, and other structures are newly constructed, disasters are completely prevented. It can be pointed out that these perceptions of people about natural disasters are affecting recognition of flood risks among the public in Japan.

Japanese flood disaster mitigation have heavily focused on preparing dams, levees, and other hardware structures. As a result, the level of frequency of occurrence of major flood disasters has declined, and the number of human casualties has fallen. However, flood-disaster preparedness measures that rely on preparation of hardware structures tend to incur large costs. Under such measures, there is also a possibility of serious damage if there is an unexpectedly heavy rainfall. In addition, as environmental problems are now attracting much public attention, a high value is set on the idea of living together with nature, making it impossible to promote public works projects that consider only disaster preparedness. From a different viewpoint, not giving the highest priority to disaster preparedness means accepting a certain level of occurrence of damage. In fact, river improvement works that presuppose certain levels of floods have been undertaken following a revision to the River Law. In line with these changes, it is necessary to come up with measures to reduce damage, instead of preparing only hardware structures in order to completely protect areas from disasters. Such measures to reduce damage include those to make houses waterproof, regulate the use of land, improve software, including information on disasters, and promote disaster preparedness measures among the public.

In recent years, much importance is being attached to integrated flood control measures that emphasize both hardware and software. Integrated flood control measures presuppose that there may be river floods. As a result, it will be more necessary than ever for residents to accept flood risks by understanding that their living areas may be flooded and to make preparations. As long as cities have been developed on alluvial plains, which have a high risk of river floods, with people and assets concentrated in such areas, it is impossible to achieve zero risk. However, as improvement in hardware is currently being promoted, residents have become totally accustomed to a situation in which there have been no flood disasters for many years. Residents may believe that levees will protect their living areas from flood disasters and perceive that zero risk can be achieved. Therefore, factors to determine consciousness to accept flood risks among residents are studied below.

2.2 Method

Participants and Procedure:

Questionnaire surveys were conducted on 4,000 households in areas that suffered from the Tokai heavy rainfall in 2000. The survey forms were mailed to 2,000 households in Nishi-ku of Nagoya-shi, 1,000 households in Shinkawa-cho, and 1,000 households in Nishibiwajima-cho, and the forms were collected through personal visits by staff. The staff asked those who failed to fill in the survey forms or those who were not at home to mail back the forms. As only some parts of Nishi-ku and Shinkawa-cho were flooded, the households in Nishi-ku and Shinkawa-cho were chosen from the flooded areas through a random sampling procedure on residential maps. As almost all parts of Nishibiwajima-cho were flooded, the households in Nishibiwajima-cho were chosen through a two-stage random sampling procedure from the residents register.

Period of surveys:

The questionnaire surveys were conducted in about 17 months after the Tokai Region Torrential Rains in 2000. The survey forms were mailed out on January 30, 2002, and the forms were collected between February 1 and February 19, 2002. The collection of the survey forms via mail was closed on February 28, 2002. Of the 4,000 households, survey forms were collected from 3,036 households (with a percentage of valid responses of 75.9%). In this research project, data from a total of 2,811 households (2,659 households living in single-family housing and 156 households living on the first floor of apartment houses) were used for analyses. The average age of the respondents was 57.6 ($SD = 12.58$).

Survey items:

There were 150 question items listed in the questionnaire. The items covered by the analyses in this section are acceptance of general risks (Yoshino and Kinoshita, 1996), consideration of society (Yoshida *et al.*, 1999), and perception regarding recognition of flood control and flood disasters, including perception of risks, consciousness about zero risk, self-responsibility, trust in administrative bodies, and interest or concern about flood disasters (refer to Appendix A). In addition, consciousness of the acceptance of flood risks is covered by the following three items: (1) it is considered to be appropriate to accept river floods to a certain extent as long as there is a risk of flood disaster in your living area; (2) there is no choice but to accept river floods to a certain extent as river floods are the works of nature; and (3) a water level up to the floor level can be tolerable to a certain extent when flood disasters occur. For all the items, responses were measured on a five-point scale ranging from 1 (disagree strongly) to 5 (agree strongly).

2.3 Results and Discussion

By carrying out structural equation modeling, a path diagram to show causal relations regarding acceptance of flood risks was created (Fig. 1). It was shown that consciousness of the acceptance of flood risks was affected by the following four factors: The first factor was consciousness of self-responsibility. The path coefficient from self-responsibility to acceptance of flood risks was .29. When people have a strong consciousness of self-responsibility, they accept flood risks. It has been revealed that people's intentions for actions regarding disaster preparedness measures against tornados or earthquakes increase when they have a strong consciousness of self-responsibility (Duval and Mulilis, 1999; Mulilis and Duval, 1995, 1997; Mulilis *et al.*, 2001). People do not carry out disaster preparedness measures unless they perceive that disasters can occur. As a result, it can be viewed that taking disaster preparedness measures is related to an attitude that is to perceive and accept disaster risks. These relationships are supported by the results of this study, and it was revealed that people accept flood risks when they have a strong consciousness of self-responsibility. Administrative bodies or experts on flood disasters are asked to try to provide information that helps increase consciousness of self-responsibility.

On the other hand, the path coefficient from consciousness of zero risk to acceptance of flood risks was -.36. In other words, residents who depend on measures to prepare dams, levees, and other hardware structures and believe that science and technology make it possible to achieve zero risk, perceive that it is impossible to accept flood risks. The information that is currently encountered by residents is about preparation of hardware structures, including improvement of levees and widening of rivers. Based on such information, the public tends to create a perception that it is possible to eliminate flood disasters. Nakayachi (2002) found that there is a low expectation for zero risk regarding natural disasters. However, the results of this study showed that many of the public seek to achieve zero risk and believe that it is possible to achieve. For example, for the question item "it is possible to achieve a society without flood disasters if public works are strengthened to carry out river improvement," 27.0% of the respondents chose 5 (agree strongly) on the five-point scale, and 32.5% chose 4 (agree moderately). This showed that there is a strong perception that it is possible to achieve zero risk. These results were

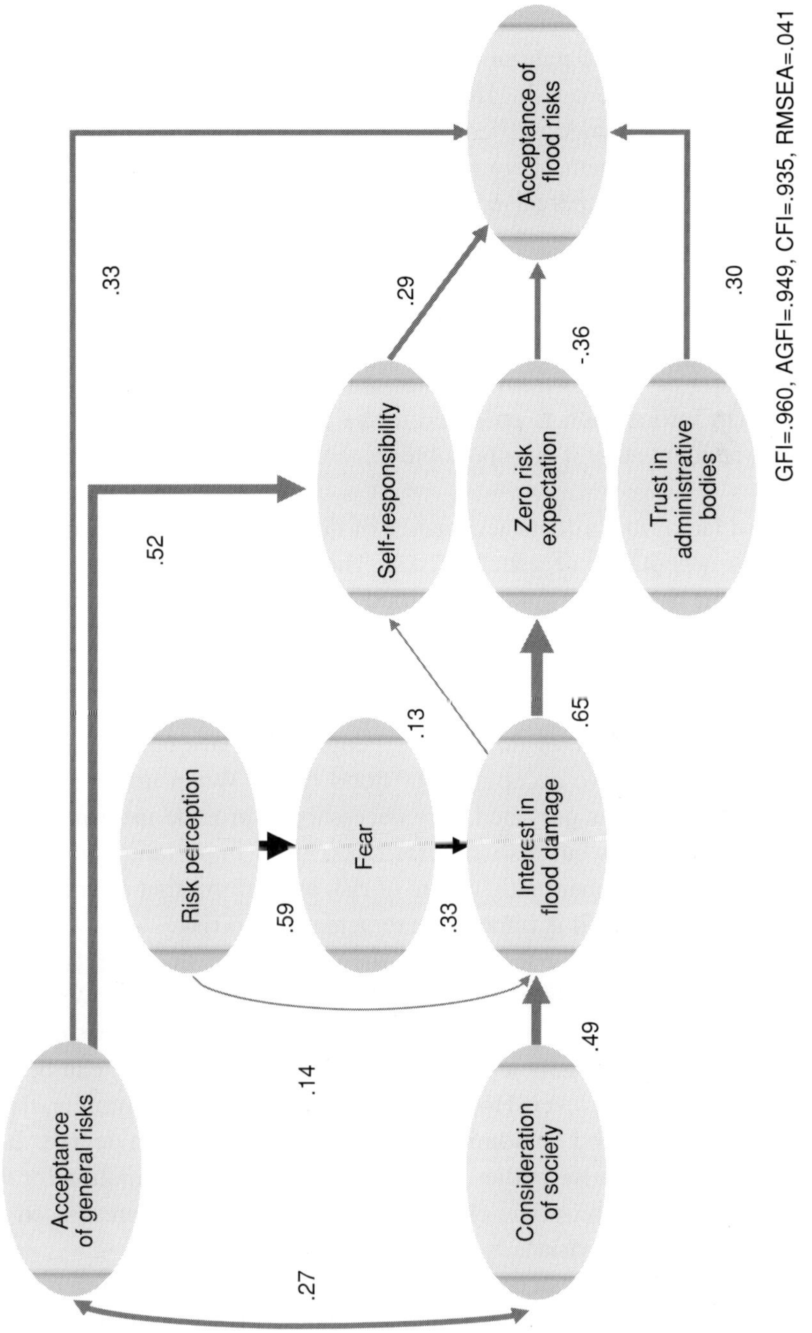

Fig. 1. Structural model of acceptance of flood risks with standardized path coefficients.

related to the "not me" factor (Joffe, 1999). In other words, people tend to perceive natural disasters as phenomena unrelated to them because there is a low occurrence probability of natural disasters. Furthermore, there is generally a low expectation for zero risk. However, when people experience a nearby disaster, become concerned parties, and recognize that natural disasters are phenomena related to them, there is a high expectation for zero risk. Experts or administrative bodies perceive as a matter of course that it is impossible to achieve zero risk regarding flood disasters, and they are asking residents to understand this by providing information through hazard maps. On the other hand, residents still have a strong consciousness that it is possible to achieve zero risk. There is a perception gap about zero risk regarding flood disasters between experts or administrative bodies and residents, who are concerned parties. When information is provided to residents, it is necessary to properly inform residents that it is impossible to achieve zero risk regarding flood disasters.

The third factor that affects acceptance of flood risks was reasonable consciousness of acceptance of general risks. The path coefficient from acceptance of general risks to acceptance of flood risks was .32. In addition, while there is a small relationship between interest in flood disasters and consciousness of self-responsibility, the path coefficient from acceptance of general risks to self-responsibility was higher at .52, confirming that there is an indirect effect on acceptance of flood risks. People who have consciousness to accept general risks, have acceptance of flood risks. When information regarding flood disasters is provided, it is necessary to provide information on not only flood disasters but also general risks. It is important also to provide the uncertain or probabilistic nature of risk and inform that all matters in society involve risks and it is impossible to achieve zero risk.

The results of this study showed that the other factor that affects acceptance of flood risks was related to trust in administrative bodies. When occurrence of flood disasters is accepted, some damage reduction measures, including buying insurance policies and making houses waterproof, can be undertaken at an individual level. However, for the whole of a community, there is no choice but to depend on administrative bodies, to a certain degree. It is necessary for administrative bodies to probide enough risk information for the residents to make an appropriate choice of risk reduction measures based on their acceptance of flood risk.

Based on the above, it can be pointed out that risk communication regarding flood disasters is necessary to promote integrated flood control measures as was illustrated in various risk scenes in Japan (e.g., Kikawa, 1999). This study showed that residents have a strong consciousness about zero risk although experts or administrative bodies are endeavoring to inform residents that it is impossible to achieve zero risk by providing information through hazard maps. In order to help residents have a proper understanding of flood risks, risk communication that enables residents and experts or administrative bodies to provide information interactively or express and exchange opinions, seems to be necessary, in addition to one-way provision of information through hazard maps and other means. In addition, risk communication between residents and administrative bodies is expected to help develop trust in the administrative bodies. The goals regarding risk communication among residents, administrative bodies, and other stakeholders, related to flood disasters include the following: (1) to strengthen consciousness about self-responsibility regarding disaster preparedness; (2) to understand that it is impossible to achieve zero risk regarding flood disasters; (3) to understanding not only flood disasters but also overall risks; and (4) to develop trust between the public and administrative bodies through risk communication.

3 Participation in Community-based Disaster Preparedness Activities

3.1 Factors to increase the intention to participate in community-based disaster preparedness activities

In Japan, where there are many natural disasters, the role of preventing disasters is partly delegated to fire-fighting organizations, flood-fighting organizations, and other community-based organizations. However, the abilities to prevent disasters in local communities are declining due to an increase in nuclear families, a decline in conventional communities, an increase in elderly people living alone, and other factors. To swiftly carry out relief activities in local communities at the initial stages after the occurrence of disasters, it is necessary to carry out community-based disaster preparedness activities on a habitual basis. In addition, it is also necessary for not only administrative bodies but also residents in local communities to participate in disaster preparedness activities, and residents need to cooperate with administrative bodies. It is desired that not only administrative bodies but also residents in local communities come up with community-based disaster-preparedness measures by themselves, which are reflected in administrative plans. To achieve this, it is necessary for more members of residents to participate in community-based disaster preparedness activities. Through past researches, it was revealed that past experience of disasters, demographic characteristics, including level of income, type of residence, and educational level, feeling of control against disasters and anxieties, which are individual characteristics, and other factors are related to disaster preparedness actions (e.g., Lindell and Perry, 2000). However, many of these researches focused on disaster preparedness actions for households, and community-based disaster preparedness activities were studied only on a supplemental basis. Among these researches, Rochford and Blocker (1991) examined disaster preparedness activities in local societies, including fund-raising campaigns, petitions to administrative bodies, and solidarity movements in local communities, and revealed that evaluations of possibilities to control flood disasters or coping styles are factors that affect participation in community-based disaster preparedness activities. However, there have been almost no findings on the level of intention to participate in community-based disaster preparedness activities among flood-fighting organizations, independent disaster preparedness organizations, and other organizations in Japan. As a result, factors to determine the intention to participate in community-based disaster preparedness activities are also studied through surveys covering residents in local areas that have a high level of weakness towards flood disasters.

Many of the past studies on disaster preparedness regarded disasters as stress events and treated disaster preparedness actions as measures to deal with such events (Mulilis and Duval, 1995; Rochford and Blocker, 1991). However, community-based disaster preparedness activities by independent disaster preparedness organizations, flood-fighting organizations, and other organizations can be regarded as voluntary activities in local societies. As a result, among the researches dealing with relationships between attitudes and actions, this research project carries out studies by using the theory of reasoned action, which is pointed out to be highly persuasive regarding voluntary actions, as the basic framework (Ajzen and Fishbein, 1977, 1980; Fishbein and Ajzen, 1975).

3.2 Method

Participants and Procedure:

A questionnaire survey was conducted to study factors that affect intention to participate in community-based disaster preparedness activities. The subjects of surveys and the period of surveys were the same as those of the surveys mentioned in the previous section.

Survey items:

As factors to determine the intention to participate in community-based disaster preparedness activities, a total of 13 items related to interest or concern about flood disasters, consciousness related to subjective norms that show expectations from important others about disaster preparedness actions, recognition of costs of community-based disaster preparedness activities, and recognition of benefits were used (refer to Appendix B). For all the items, responses were measured on a five-point scale ranging from 1 (disagree strongly) to 5 (agree strongly).

3.3 Results and Discussion

As to the condition of participation in disaster preparedness classes or disaster preparedness drills, 188 respondents (6.2%) said that they attend all such events, 1,330 (43.8%) said that they sometimes attend such events, 658 (21.7%) said they cannot attend such events because they cannot obtain information, 550 (18.1%) said that they do not attend such events because they have no interest, 183 (6.0%) said that they cannot attend such events because they are disabled or sick, and 127 (4.2%) did not respond. Almost half of the respondents said that they have experience in participating in community-based disaster preparedness activities. Data from 2,726 respondents, excluding 310 respondents who said that they cannot attend disaster preparedness classes or drills because they are disabled or sick and who did not respond, were used for the following analyses.

By carrying out structural equation modeling, a path diagram to show causal relations regarding participation in community-based disaster preparedness activities was created (Fig. 2). From the results of the analyses, it was revealed that people have a high intention to participate in community-based disaster preparedness activities when they have a high intention related to a subjective norm. Subjective norm is an important determinant of regarding public activities carried out in the presence of acquainted others (Hirose, 1992). These estimates are supported by the results of this research project, which suggest that a subjective norm is an important factor to determine intention regarding actions carried out in the presence of others.

In addition, recognition of costs has a strong negative effect on intention to participate. When people have a high recognition of costs, their intention to participate declines. On the other hand, recognition of benefits has only a small positive effect on intention to participate. As disasters do not occur frequently, people feel highly burdened to participate in community-based disaster preparedness activities during normal times when nothing happens. To activate community-based disaster preparedness activities, it is extremely important to reduce the public's recognition of costs. On the other hand, recognition of benefits had only a small effect. Damages from flood disasters cannot be eliminated only through community-based disaster preparedness activities. Furthermore, it is difficult to see detailed benefits from carrying out community-based disaster preparedness activities. It could be pointed out that these are the reasons why recognition of benefits had only a small effect.

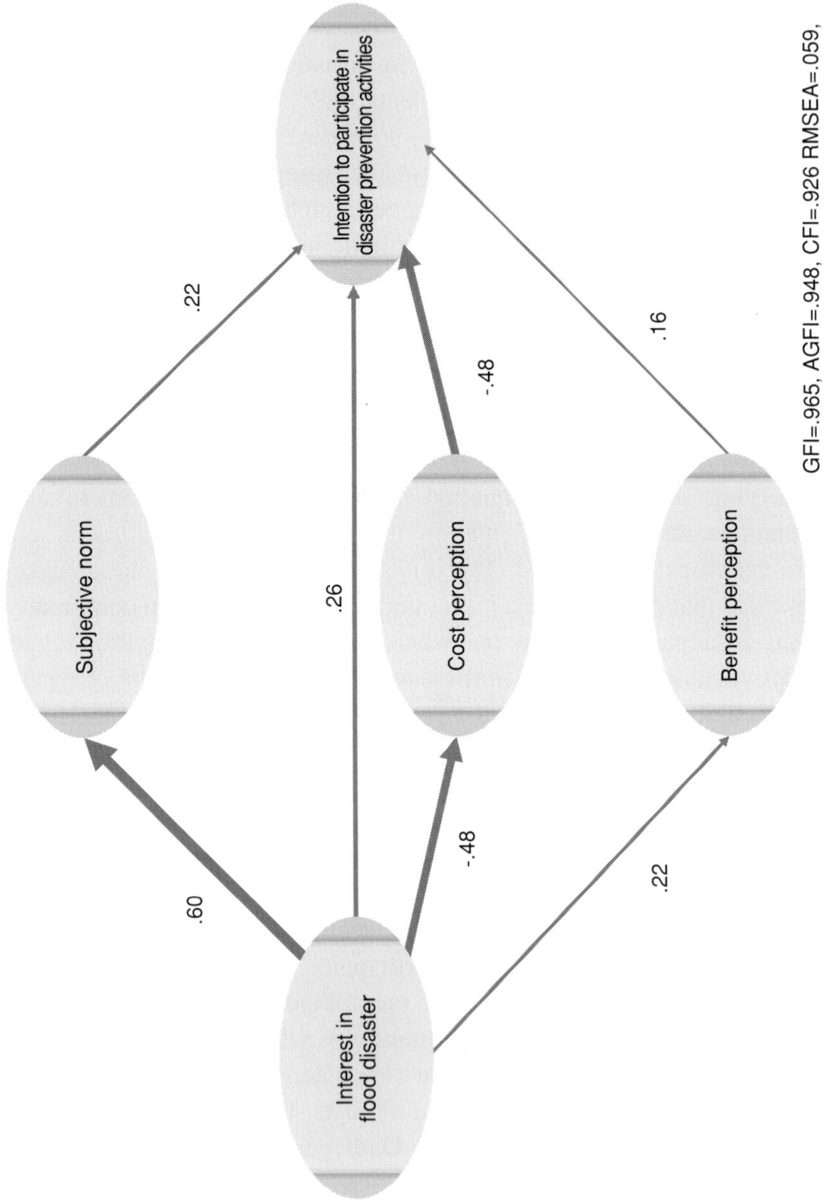

Fig. 2. Structural model of intention to participate in disaster prevention activities with standardized path coefficients.

Interest in or concern about flood disasters is a conception that is not dealt with under the theory of reasoned action. However, interest in or concern about flood disasters serves as a direct factor that determines intention regarding actions in this study. In addition, interest or concern about flood disasters was regarded as a factor that heightens a subjective norm or recognition of benefits and lowers recognition of costs. Furthermore, interest in or concern about flood disasters had an indirect effect on intention regarding actions. Interest or concern about disasters serves as an important factor to increase intention regarding community-based disaster preparedness actions. To activate community-based disaster preparedness activities, it is first important to encourage residents to take interest in disasters in local areas. It was suggested that it is possible to activate community-based disaster preparedness activities by providing information in such a way that residents takes more interest in disasters.

Research that studied relationships between social attributes and disaster preparedness measures (Russell *et al.*, 1995), research that studied disaster preparedness activities as stress-coping activities by regarding disasters as stress events (Rochford and Blocker, 1991), and other researches have been carried out in the past. On the other hand, this study regarded disaster preparedness activities as voluntary activities in local areas and revealed that the three factors—subjective norms, recognition of costs, and interest or concern about disaster preparedness—affect intention regarding community-based disaster preparedness actions. This points out that it is possible to activate community-based disaster preparedness activities by regarding disaster preparedness activities not as measures to reduce disaster risks but as voluntary activities in local communities. Disaster preparedness measures are often studied from the viewpoint of completely understanding risk information and making preparations. However, it is also necessary to carry out studies to activate community-based disaster preparedness activities from the viewpoint of focusing on community-based activities.

4 Conclusion

Two findings were obtained through this study. First, self-responsibility, consciousness about zero risk, acceptance of ordinary risks, and trust in administrative bodies affect acceptance of flood risks. From this result, it is pointed out that in order to promote acceptance of flood risks, it is important to increase self-consciousness about disaster preparedness through risk communication, understand that it is impossible to achieve zero risk, deepen understanding of not only flood disasters but also overall risks, and establish relationships of trust between the public and administrative bodies.

Second, people have a greater intention to participate in community-based disaster preparedness activities when they take great interest in subjective norms and flood disasters. Furthermore, recognition of costs of disaster preparedness activities serves as a factor to decrease intention to participate. As a result, it was pointed out that it is possible to activate disaster preparedness activities by regarding these activities not as measures to reduce disaster risks but as voluntary activities in local communities.

Appendix A. Questionnaire survey items in Section 2

Reasonable perception of general risk

- I think everything that occurs is accompanied by risk, and nothing can be done about it.
- I think that the world is made up of risk and safety.
- I think living with risk is a fact of life.

Consideration of society

- I think about how society is made up.
- I think about how I should act in society.
- I think about the society in which I live.

Appendix B. Questionnaire survey items in Section 3

Interest or concern about flood disasters

- I feel concern for the details of flood damage mitigation measures being taken by administrative organizations.
- I often read newspaper articles about flood damage.
- I am interested in how much money they spend for flood damage mitigation measures and flood control works.

Subjective norms

- I will feel ashamed if I make no preparation while my relatives and family were taking action for flood protection.
- If I take action for flood protection, I think my close friends will be impressed with what I do.
- I will feel ashamed if I do nothing while my neighbors are taking measures for flood protection.

Cost and benefit perception

- I think it's difficult to find the time to participate in evacuation drills and disaster prevention seminars in the local community.
- I think it's too much of a bother to check if there are any flood-prone areas in residential areas.
- I think that flood damage can be minimized if everyone takes disaster prevention measures.
- I don't think whether or not I do something to reduce flood damage makes any difference.

Intention to participate in community based disaster preparedness

- I want to participate in disaster preparedness classes or disaster preparedness drills.
- If administrative organizations or fire stations hold seminars or disaster prevention drills, I want to participate.
- I want to join a voluntary disaster prevention organization

Participation in local disaster prevention activities

- How often do you participate in disaster preparedness classes or disaster preparedness drills?

References

Ajzen, I. and M. Fishbein (1977): Attitude-behavior relations: A theoretical analysis and review of empirical research, *Psychological Bulletin*, **84**, 888–918.

Ajzen, I. and M. Fishbein (1980): Understanding attitudes and predicting social behavior, Englewood Cliffs, NJ, Prentice-Halll.

Duval, T. S. and J.-P. Mulilis (1999): A person-relative-to-event (PrE) approach to negative threat appeals: A field study, *Journal of Applied Social Psychology*, **29**, 495–516.

Fishbein, M. and I. Ajzen (1975): Belief, attitude, intention, and behavior: An introduction to theory and research, Readings, MA, Addison Wesley.

Hirose., Y. (1992): Decision Making Model of Environmental Consumer Behavior: Reappraisal of Fishbein and Ajzen's Attitude Model, *Japanese Psychological Review*, **35**, 339–360 (in Japanese).

Joffe, H. (1999): Risk and 'the Other.' Cambridge, Cambridge University Press.

Kikkawa, T. (1999): Risk Communication, Fukumura Shuppan Inc (in Japanese).

Lindell, M. K. and R. W. Perry (2000): Household adjustment to earthquake hazard: A review of research, *Environment & Behavior*, **32**, 461–501.

Mulilis, J.-P. and T. S. Duval (1995): Negative threat appeals and earthquake preparedness: A person-relative-to-Event (PrE) model of coping with threat, *Journal of Applied Social Psychology*, **25**, 1319–1339.

Mulilis, J.-P. and T. S. Duval (1997): The PrE model of coping and tornado preparedness: Moderating effects of responsibility, *Journal of Applied Social Psychology*, **27**, 1750–1766.

Mulilis, J.-P., T. S. Duval, and D. Rombach (2001): Personal responsibility for tornado preparedness: Commitment or choice, *Journal of Applied Social Psychology*, **31**, 1659–1688.

Nakayachi, K. (2002): Classification of the desire for zero risk attainment: A exploratory study of tis cognitive characteristics, *Japanese Journal of Social Psychology*, **17**, 63–72 (in Japanese).

Rochford, E. B. and T. J. Blocker (1991): Coping with "natural" hazards as stressors: The predictors of activism in a flood disaster, *Environment and Behavior*, **23**, 171–194.

Russell, L., J. D. Goltz, and L. B. Bourque (1995): Preparedness and hazard mitigation actions before and after two earthquakes, *Environment and Behavior*, **27**, 744–770.

Slovic, P., H. Kunreuther, and G. F. White (1974): Decision processes, rationality and adjustment to natural hazards, edited by G. F. White, Natural hazards, local, national, and global, New York, Oxford University Press, pp. 187–205.

Yoshida, T., N. Ando, T. Motoyoshi, T. Fujita, S. Hirooka, K. Saito, K. Mori, Y. Isida, and M. Kitaori (1999): A study of Social Annoyance I, *Bulletin of the School of Education: Psychology*, **46**, 53–73 (in Japanese).

Yoshino, K. and T. Kinoshita (1996): Constructing the "Scale of Risk Acceptance: SRA", *Japanese Journal of Risk Analysis*, **9**, 121–124 (in Japanese).

A better integrated management of disaster risks: Toward resilient society to emerging disaster risks in mega-cities, Eds., S. Ikeda, T. Fukuzono, and T. Sato, pp. 135–151.

Residents' Perception about Disaster Prevention and Action for Risk Mitigation: The case of the Tokai flood in 2000

Kenji Takao

1 Introduction

The Tokai flood struck in 2000 in Nagoya City and adjacent areas. The affected areas were near the Shinkawa River and areas where levees were prone to overflow. One might think that such damage would not have happened if the embankment or pumping functions had been improved before the flood. Traditional measures to mitigate the effect of floods have included building dams, reservoirs, and embankments. Although these so-called 'hard' measures may seem effective, in reality they do not always prevent flood damage (Burton *et al.*, 1993). Hard measures are based on various calculations, such as the likelihood of heavy rainfall once every 100 years, which might cause damage greater than some pre-calculated value. In addition, hard measures require considerable amounts of money and construction resources. Therefore, both hard and soft measures, the latter of which raises the level of residents' preparedness against floods, must be considered. In other words, residents' preparedness against natural disasters is important.

How many people prepare against natural disasters? Past research has suggested that residents' preparedness against natural disasters is not enough for preventing natural disasters. Mizuno (1981) revealed, through a questionnaire survey with a multiple-selection method, that the percentage of residents who participated in disaster prevention activities was only 16.7%. Furthermore, those who conformed to the evacuation center in their region amounted to 31.7%. Hashimoto *et al.* (2001) showed that 19% of residents in the affected area perceived that their preparedness for evacuating due to disaster was insufficient. In addition, 78% residents in the area did not prepare at all for disaster. Despite the need for residents to be prepared, many do not prepare at all.

What kind of factors affect residents' preparedness against flood? This chapter focuses on the influence of psychological factors and home ownership in determining preparedness against floods, factors that are most often neglected by social psychologists.

2 Why are people unwilling to prepare for floods?

Some studies have found a link between home ownership and preparedness for disasters. Mulilis *et al.* (2000) showed that home owners prepared for tornados more thoroughly than did people living in rented accommodation. They concluded that home owners valued their property much more highly than the resources needed to prepare for floods; these home owners felt more compelled to prepare for tornados than did those who lived in rented accommodation. These findings suggest that flood anticipation and experience may not necessarily determine a person's preparedness for floods; rather, home ownership determines preparedness for disasters.

To improve residents' preparedness for floods, the source of this lack of correspondence between flood related perception and preparedness must be examined. However, little has been done to establish a statistically quantitative relationship between flood related perception and preparedness for floods. The impact of home ownership on preparedness for floods is also unclear. Quantitatively identifying some determinants of people's preparedness for floods may facilitate the promotion of flood risk management. Takao *et al.* (2004) identified four goals in establishing the factors that determine a person's preparedness for floods. The first goal was to examine the relationship between flood related perception (the level of fear of floods and flood anticipation) and preparedness. The second was to investigate the relationship between flood experience and preparedness. The third was to analyze the relationship between the amount of damage from a previous flood and preparedness. The fourth goal was to examine how owning a home affected a person's preparedness for floods. The next section contains a review of these goals described in Takao *et al.* (2004).

3 Questionnaire Survey

3.1 Method

a) Survey area and subjects

A questionnaire-based survey was conducted in Nagoya City's Nishi Ward and in Shinkawa Town (Fig. 1), areas affected by the Tokai flood. The flood claimed several lives; four houses completely collapsed, 98 houses partially collapsed, 11,142 houses were flooded above floor level and 21,852 were flooded below floor level. In Shinkawa Town, 2,391 houses were flooded above floor level and 1,244 below floor level. Questionnaires were sent to 5,979 residents of Nagoya City and Shinkawa Town. The questionnaires were not received by 235 people. Of the remaining 5,744 people, 2,051 (35.7%) responded.

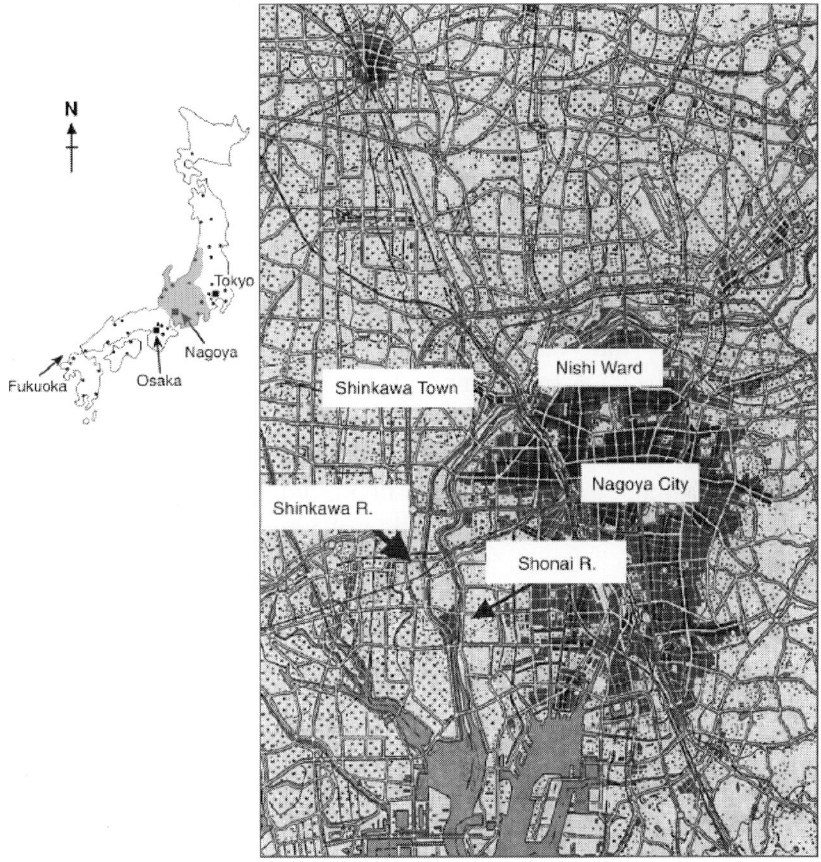

Fig. 1. Map of the surveyed area.

b) Survey methodology

Preparedness for floods was examined in terms of measures taken by the residents after the Tokai flood. Measures taken before include taking out insurance and checking the hazard map of the area, while measures taken after means taking special measures for future floods. The chi-square test was conducted to examine the relationship between these measures and such factors as flood experience, fear of floods, flood anticipation, and flood risk perception.

The respondents rated their flood experience by answering the question "Had you ever experienced a flood before the Tokai flood?". They were also asked to rate their flood related perception in terms of flood anticipation and their fear of floods. The respondents were asked to rate their flood anticipation using a two-point scale (yes or no) to respond to the question "Before the Tokai flood, did you ever think that you may have another flood in your area?". In this study, term "flood anticipation" was based not on some objective criteria such as the cumulative amount of rain, but on the personal judgment of the respondents. They were also asked to rate their level of fear of floods using a three-point scale (very fearful, fearful, not so fearful). The respondents were also asked to rate the amount of damage sustained in the Tokai flood using a four-point scale (complete or partial collapse, flooded above floor level, flooded below floor level, no damage). They were asked about their level of preparedness: "Had you taken out an insurance policy, checked the hazard map of the area?". A two-point scale was used and the respondents were asked to check the measures they had undertaken before and after the Tokai flood to prepare for floods. They were also asked about their preparedness after the Tokai flood: "After the Tokai flood, did you take measures against future floods?". Again, a two-point scale was used. They were also asked to state whether they owned their home or lived in rented accommodation.

3.2 Results

a) Determinant factors of flood preparedness (Fig. 2)

Preparedness before the Tokai flood: taking out an insurance policy and checking the hazard map

The relationship between flood experience and residents' preparation for floods was analysed. It was found that neither taking out an insurance policy nor checking the hazard map was associated with flood experience (taking out an insurance policy: $\chi^2(1) = 2.72$, n.s.; checking the hazard map: $\chi^2(1) = 1.51$, n.s.). This indicates that flood experience was not related to preparedness for floods. The relationship between risk perception, fear of floods and preparedness for floods was also analysed. The analysis showed that taking out an insurance policy ($\chi^2(1) = 1.47$, n.s.) and checking the hazard map ($\chi^2(1) = 0.27$, n.s.) were not associated with flood anticipation. Further analyses illustrated that taking out an insurance policy was associated with fear of floods (taking out an insurance policy: $\chi^2(2) = 6.70$, $p < 0.05$; Fig. 2(a)). The relationship between the residents' fear of floods and whether or not they checked the hazard map, however, was not significant ($\chi^2(2) = 0.01$, n.s.). These findings indicate that those who were more fearful of floods were more likely to take out an insurance policy. In summary, there was no relationship between the residents' preparedness for floods and their anticipation of floods.

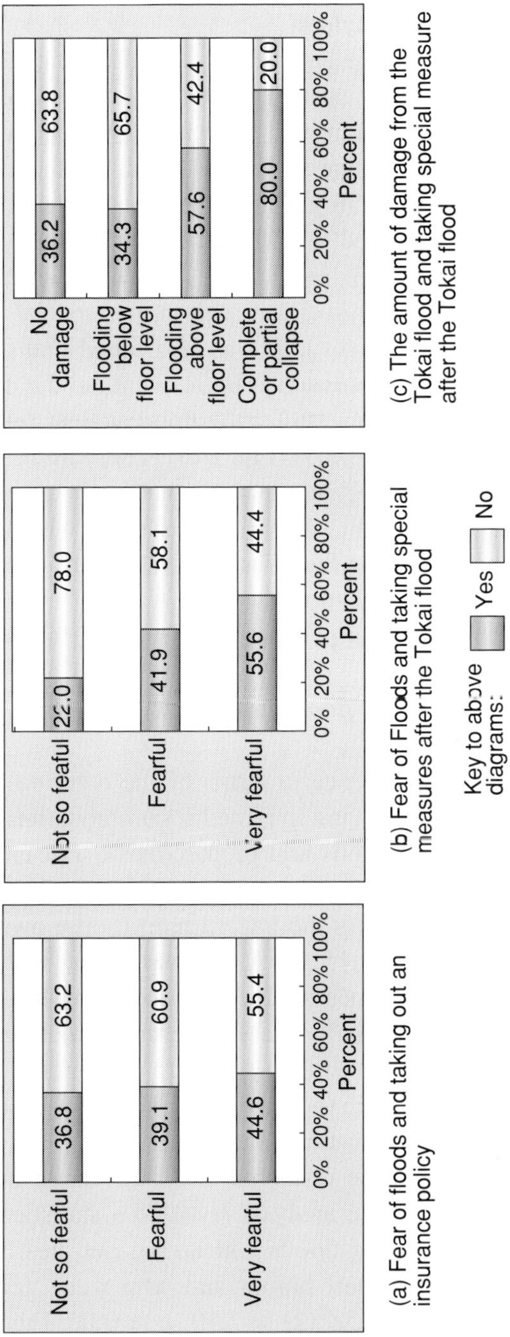

Fig. 2. The determinant factors of preparedness for floods.

Preparedness after the Tokai flood: flood anticipation, fear of floods, and amount of damage sustained in flood

We analysed the relationship between flood anticipation, fear of floods, and whether or not residents took special measures after the Tokai flood against future floods. In addition, we examined the relationship between the amount of damage sustained by the residents in the Tokai flood and whether they took special measures after the Tokai flood against future floods. The fear of floods was found to be related to taking special measures after the Tokai flood against future floods ($\chi^2(3) = 66.58$, $p < 0.01$) (Fig. 2(b)). However, anticipation of floods was not associated with taking special measures after the Tokai flood ($\chi^2(1) = 1.76$, n.s.). It was also found that the relationship between the amount of damage sustained by residents in the Tokai flood and whether they took special measures after the flood against future floods was significant ($\chi^2(3) = 74.52$, $p < 0.01$) (Fig. 2(c)). These findings indicate that those who feared floods to be high were more likely to take special measures against future floods. These results also show that those who had suffered serious damage from the Tokai flood were more likely to take special measures against floods than those who had not.

b) Determinant factors of flood preparedness: the difference in terms of home ownership (Fig. 3)

Preparedness before the Tokai flood: taking out an insurance policy and checking the hazard map

There was almost no difference in terms of the determinant factors between home owners and those living in rented accommodation, except for the relationship between fear of floods and preparedness. For example, neither taking out an insurance policy (home owners: $\chi^2(1) = 2.81$, n.s.; renters: $\chi^2(1) = 0.25$, n.s.) nor checking the hazard map (home owners: $\chi^2(1) = 0.17$, n.s.; renters: $\chi^2(1) = 3.15$, n.s.) was associated with flood experience. Thus, home ownership does not affect the relationship between flood experience and preparedness for floods. Furthermore, neither taking out an insurance policy (home owners: $\chi^2(1) = 0.22$, n.s.; renters: $\chi^2(1) = 0.03$, n.s.) nor checking the hazard map (home owners: $\chi^2(1) = 0.40$, n.s.; renters: $\chi^2(1) = 0.01$, n.s.) was related to flood anticipation.

The relationship between the residents' level of fear of floods and their preparedness was analysed. The analysis revealed a statistically significant relationship between the fear of floods and taking out an insurance policy for the residents who owned their homes and who were 'fearful' or 'very fearful' of floods (home owners: $\chi^2(2) = 7.10$, $p < 0.05$; Fig. 3(a)) (renters: $\chi^2(2) = 0.18$, n.s.; Fig. 3(b)). However, there was no relationship between

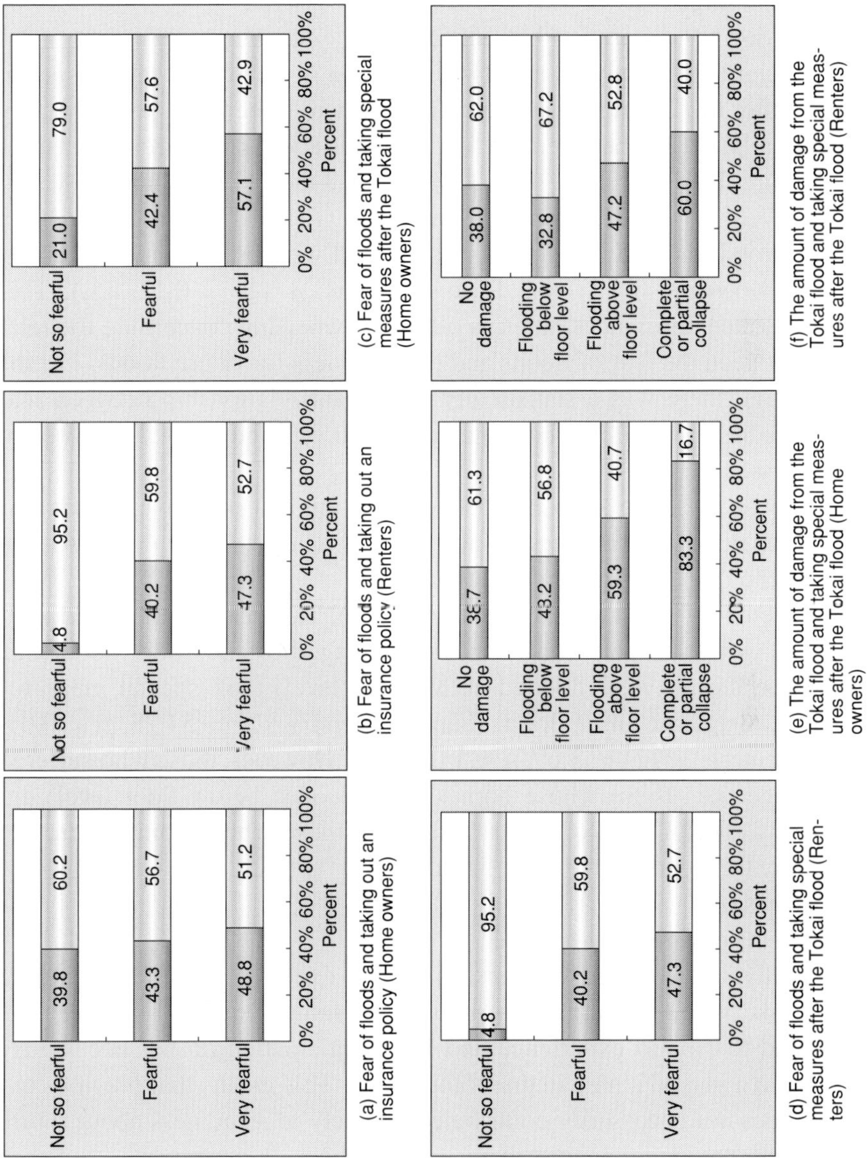

Fig. 3. The determinant factors of preparedness for floods: the difference in term of home ownership.

the fear of floods and checking the hazard map (home owners: $\chi^2(2) = 0.05$, n.s.; renters: $\chi^2(2) = 0.53$, n.s.).

c) Preparedness after the Tokai flood: flood anticipation, fear of floods, and amount of damage

In addition to preparation beforehand, for home owners there was a statistically significant relationship between the fear of floods and taking special measures after the Tokai flood against future floods. There was a statistically significant relationship between the fear of floods and taking special measures against future floods for the residents who owned their homes (home owners: $\chi^2(2) = 67.14$, $p < 0.01$, Fig. 3(c); renters: $\chi^2(2) = 1.16$, n.s., Fig. 3(d)). However, flood anticipation was not associated with preparedness after the flood (home owners: $\chi^2(1) = 2.18$, n.s.; renters: $\chi^2(1) = 0.17$, n.s.). This finding indicates differences in terms of home ownership concerning the relationship between the fear of floods and preparedness for future floods. For the home owners, there was a statistically significant relationship between fear and preparedness, while for the renters, the fear of floods was not related to preparedness.

There was also a relationship between the amount of damage sustained in the Tokai flood and taking special measures against future floods for the residents who owned their homes (home owners: $\chi^2(2) = 54.28$, $p < 0.01$, Fig. 3(e); renters: $\chi^2(2) = 3.69$, n.s., Fig. 3(f)). Residents who owned their homes and whose property was seriously damaged by the Tokai flood (i.e., those whose homes were flooded above floor level) took special measures after the flood against future floods (home owners: $\chi^2(2) = 54.28$, $p < 0.01$, Fig. 3(e); renters: $\chi^2(2) = 3.69$, n.s., Fig. 3(f)) . However, those who suffered little damage (i.e., those whose homes were flooded below floor level) did not take any special measures and neither did those whose homes were not damaged by the flood.

3.3 Discussion

a) The effects of flood experience

Generally, it is believed that those who have experienced a flood are more likely to take special measures to prepare for future floods. However, this study found that flood experience and flood anticipation do not necessarily contribute to a person's preparedness for floods. This means that previous experience does not necessarily contribute positively to awareness about future floods and preparedness against them.

b) Effects of flood anticipation and fear of floods

There was a statistically significant relationship between the fear of floods and preparedness for floods, although flood anticipation was not related to preparedness. It was also found that home ownership affected the residents' preparedness for floods and the relationship between the fear of floods and preparedness. Those home owners who were more fearful of floods were more likely to prepare for future floods than those who were less fearful of floods. In contrast, those who lived in rented accommodation did not necessarily improve their preparedness for floods, even if they were fearful of floods. This result suggests that emotional response has a stronger effect on preparedness for floods than cognitive response such as flood anticipation.

Sjöberg (1998) was the first to suggest the difference between emotional response and cognitive response in events that involve a certain amount of risk, such as traffic accidents or thunderstorms. Assuming the validity of these findings, it can be concluded, based on the results of this study, that cognitive response (e.g., flood anticipation) does not contribute to residents' preparedness for floods; rather it is the emotional response (e.g., fear of floods) that is strongly related to residents' preparedness. The results of the study are, however, insufficient to permit conclusions about the difference between emotional and cognitive reactions.

c) The effect of home ownership

A crucial factor found in this study to have determined preparedness for floods was home ownership. Thus Mulilis *et al.*'s (2000) conclusion that home owners feel more in control of their lives and tend to be more responsible than renters with respect to preparedness against tornados should not be surprising. A similar conclusion can be drawn in the case of floods. For home owners, there was a strong relationship between the fear of floods and taking out an insurance policy as well as taking special measures after the Tokai flood against future floods. In addition, there was a statistically significant relationship between the amount of damage in the Tokai flood and preparedness for floods among home owners. These results mean that residents who own their homes manage the risk of floods based on the amount of damage from previous floods. In contrast, the fear of floods and the amount of damage sustained in a previous flood by renters do not necessarily contribute to their preparedness for future floods. This is because those who own their homes are more likely than renters to want to protect their property from floods. This result implies that home owners invest more time and money in constructing their homes and acquiring household goods than those who live in rented accommodation. Therefore, home owners are likely to place more emphasis on their preparedness for floods and are likely to be more fearful of floods than renters.

4 Do flood risk perception and fear of floods directly affect preparedness against floods?

4.1 Hypothesis model of residents' preparedness against floods

It is generally believed that flood-risk perception and flood experience are directly related to preparedness against floods. Our findings, however, indicate that these do not always directly affect residents' preparedness. This suggests that other factors are related to the behavioral intention to prepare against floods. Therefore, we need to reconsider the determinant factors of residents' preparedness against flood in this chapter.

A prior psychological study categorized worry as an emotional response and risk perception as a cognitive one in events such as thunderstorms (Sjöberg, 1998). Those who perceive flood risk and fear of floods will try to reduce their worries because they feel uneasy in dangerous situations. If this hypothesis is right, both flood-risk perception and fear of floods should increase individuals' sense of self-responsibility and concern about floods because residents who perceive flood risk and who are afraid of floods, will voluntarily learn about floods or preparedness.

Both concerns about floods and the sense of self-responsibility should affect preparedness against floods. The sense of self-responsibility should increase the concern about floods because those who have the sense of self-responsibility need to check potentially dangerous areas on flood maps and taking out insurance policies themselves. Based on that thinking, we can predict that the sense of self-responsibility and the concerns about floods are related to the behavioral intentions about floods. Furthermore, the sense of self-responsibility should be related to the concern about floods. Such factors might affect residents' preparedness against floods.

Moreover, the residents' evaluations of flood policies are related to their own preparedness against floods. When the public administrators decide on flood prevention policies without explaining these to residents, the residents may not voluntarily accept their decisions. Prior social psychological studies have revealed that both those who perceived that public administrators listened to residents' opinions and those who perceived that they had explained their decisions were willing to accept these decisions (Takao, 2002; Takao *et al.*, 2003a; Tyler and Lind, 1992; Tyler *et al.*, 1997). In other words, those who evaluated the decision-making process for policies as being fair were willing to obey the decisions made by public administrators, including being prepared for floods (see hypothesis model in Fig. 4). The next section contains an analysis of the hypothesis model.

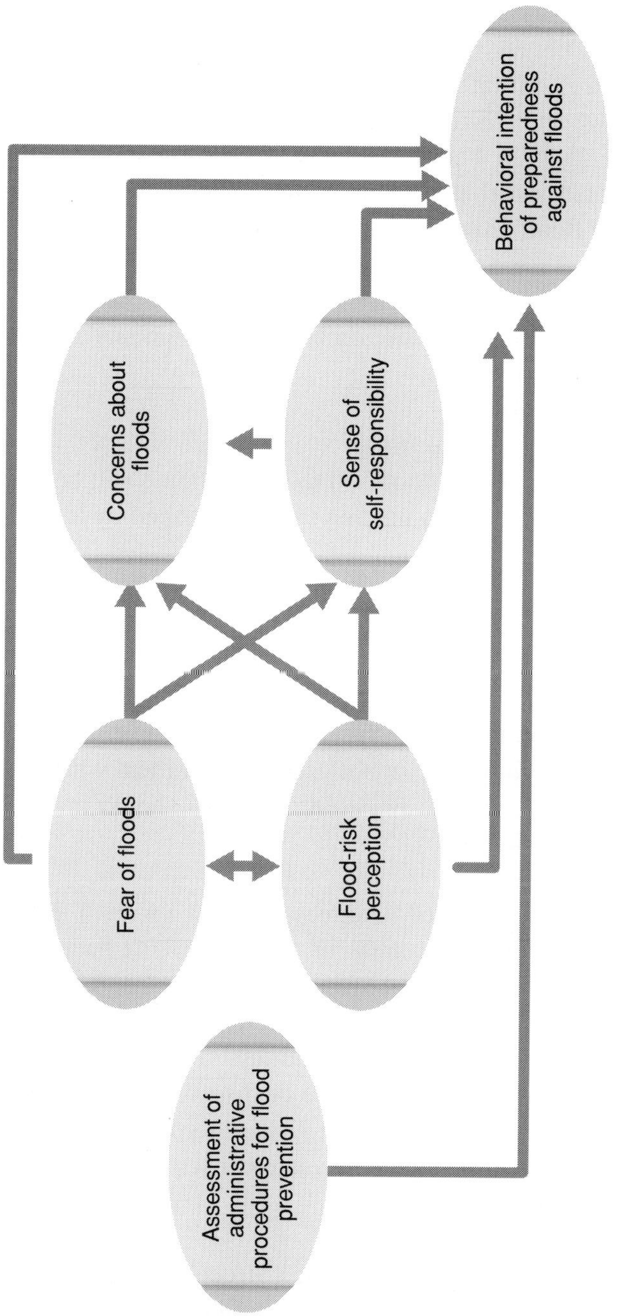

Fig. 4. Hypothesis model (Takao *et al.*, 2003b).

4.2 Methodology

A questionnaire-based survey was conducted in Nagoya City and adjacent areas (Fig. 1) in February 2002, 17 months after the disaster occurred. We mailed the questionaire to 4,000 residents of Nagoya City and the adjacent areas of Shinkawa and Nishibiwajima Town. We received 3,036 replies (i.e., response rate of 75.9%). The following items were assessed using a 5-point Likert scale that ranged from "I strongly disagree" to "I strongly agree".

a) Evaluation of flood administrative procedures

The respondents were asked to agree or disagree with the statements: "The administration explained their decisions to residents in making flood prevention policies", and "The administration listened to residents' opinions in making flood prevention policies".

b) Flood risk perception

The respondents were asked to agree or disagree with "I think my home is capable of being damaged by flood", "I think my home will be flooded above floor level", and "I think my living area can be damaged by flood".

c) Fear of floods

The respondents were asked to agree or disagree with "I feel a fear of floods when it rains" and "When I think about rain, I feel uncomfortable".

d) Sense of self-responsibility

The respondents were asked to agree or disagree with "Residents should prepare emergency rations by themselves", "Residents should prepare for floods because public agencies cannot protect all of them when unexpectedly heavy rainfalls occur", and "If possible, residents should protect both their household goods and their lives".

e) Concern about floods

The respondents were asked to agree or disagree with "I think we should check flood maps for potential areas of inundation", "I had concerns about taking the measures recommended by public agencies", and "Floods represent a serious threat for me".

f) Behavioral intention of flood preparedness

The respondents were asked to agree or disagree with "I want to participate in a residents' flood action group", "I want to participate in disaster prevention activities in my town", and "I want to participate in disaster prevention seminars or disaster prevention training".

4.3 Results

We used structural equation modeling. Structural equation modeling is a statistical technique commonly used in questionnaire-based surveys to establish the causal relationship among variables. The values are standardized path coefficients. The model was estimated using the covariance structure analysis in AMOS (Fig. 5). The results indicated that flood-risk perceptions, fear of floods, and sense of self-responsibility were not directly related to behavioral intention of preparedness against floods. The analysis also indicated that the fear of floods was not related to self-responsibility. Furthermore, flood risk perceptions were not related to concerns about floods. We modified the hypothesis model based on these results to create the one in Fig. 6.

The chi-square test revealed that the model was statistically significant. Generally, when there are more than a thousand sample models, chi-square tests indicate they are statistically significant. We also estimated its goodness of fit index (GFI), adjusted goodness of fit index (AGFI), confirmatory factor index (CFI), and root mean square error of approximation (RMSEA) in this study. Covariance structure analysis was conducted on the modified model. Values above 0.90 on the GFI, AGFI, and CFI are generally considered a good fit, as are values below 0.08 on the RMSEA.

The modified model yielded values of 0.953 on the GFI, 0.936 on the AGFI, 0.915 on the CFI, and 0.061 on the RMSEA and analysis indicated that the model was valid. The modified model indicated a strong correlation between flood-risk perceptions and the fear of floods. Flood-risk perceptions are strongly related to the sense of self-responsibility. Furthermore, the fear of floods is related to concerns about floods. The sense of self-responsibility is not strongly affected by assessments of administrative procedure on flood policies and the fear of floods. Thus, the sense of self-responsibility is determined mainly by flood-risk perceptions. The sense of self-responsibility determines concern about floods. Also, concern about floods is related to the intention to prepare against floods.

4.4 Discussion

These results mean that both the flood-risk perception and the fear of floods do not directly affect personal preparedness against floods. The perception of flood risk was strongly related to residents' sense of self-responsibility in being prepared against floods, but it was not related to their willingness to know more about them. However, their fear of floods was related to actual concerns about them; this was an important factor that affected their attitudes toward preparedness. This means that the sense of self-responsibility and the fear of floods were important factors that affected their concerns about floods. The results also indicated that residents' willingness to learn more about floods was an important factor that affected their attitudes toward preparedness.

Public administrators are generally trying to promote residents' aware-

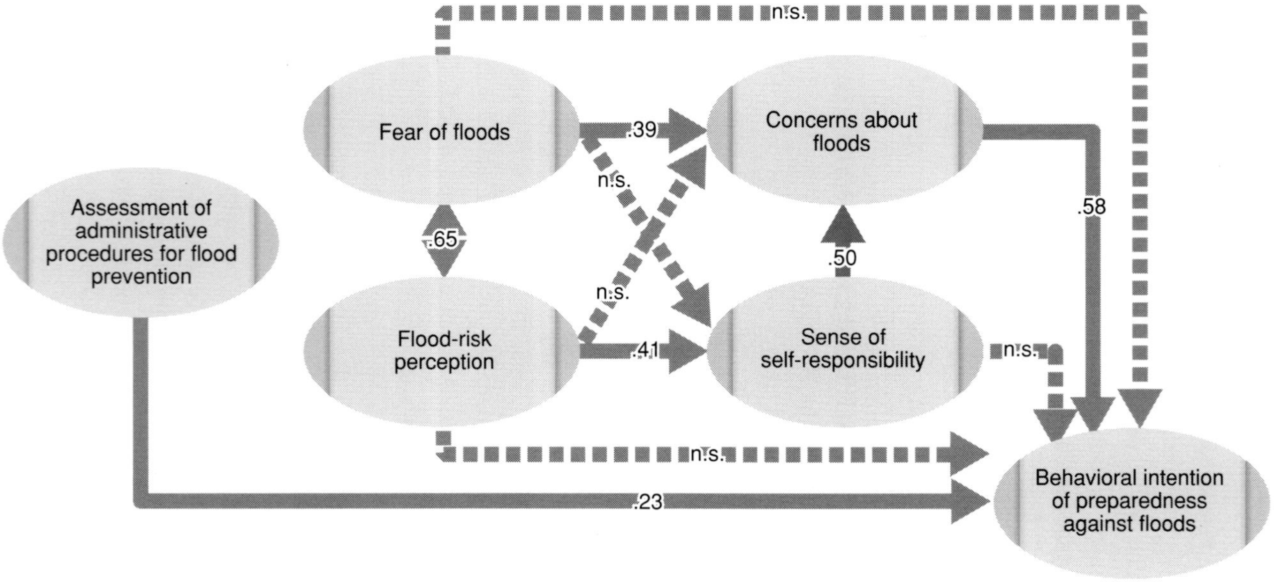

Fig. 5. Analyzed model.
Note: The values are standardized path coefficient. The sign of "n.s." means that there is not statistically significant relationship.

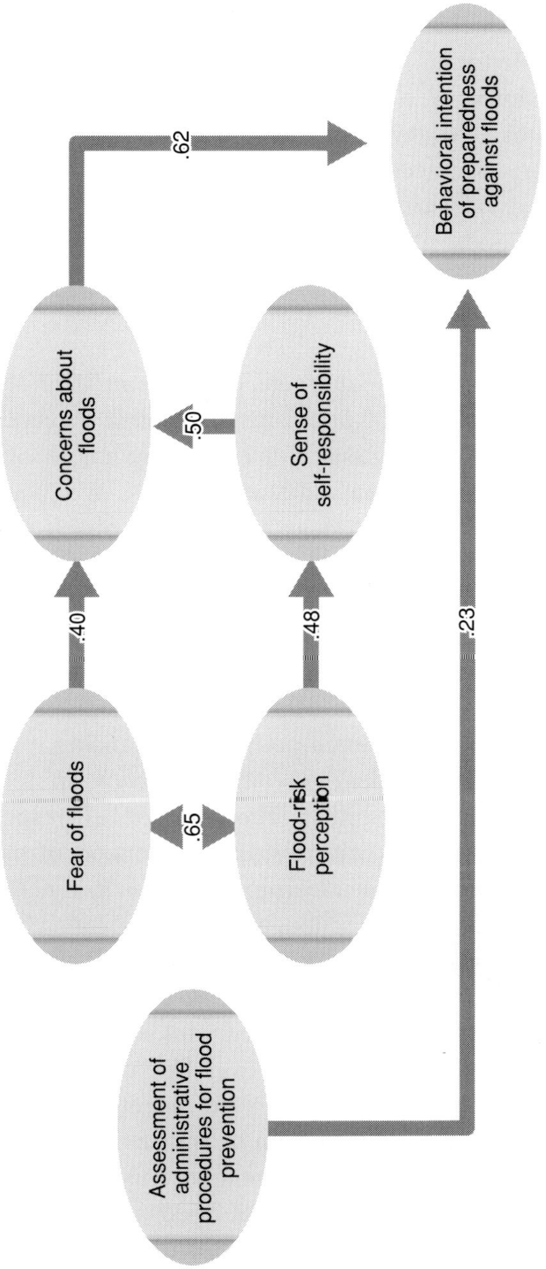

Fig. 6. Modified model.
Note: The values are standardized path coefficient.

ness of self-responsibility (e.g., Aichi Prefectural Government, 2000; Cabinet Office, 2002) because they cannot always protect all residents in time of unexpectedly heavy rainfalls. This study discovered that if residents were not individually responsible for their own preparedness, they did not demonstrate concern for floods. Moreover, if residents did not fear floods, they were not concerned about floods. We therefore need to construct a learning system whereby residents can have access to issues related to floods. The amount of information we give to residents should depend on their perceived level of flood risk and fear of floods. The system would motivate residents to participate in resident-based risk management systems.

5 Concluding Remarks

Our analysis shows that we should focus on residents' fear of floods and that the fact that residents' fear of floods and flood-risk perception are not directly related to preparedness against floods. It may not be right to value only residents' flood-risk perception and fear of floods. We should also pay attention to some psychological factors such as a sense of self-responsibility and concern with floods. Almost all learning systems for flood prevention for residents, however, have been focused on the sense of risk and fear of floods. Therefore, it is important that residents should learn about floods through a new education system that takes into account the models in this study.

However, this will not be easy to achieve. How should we do it? The above analyses suggest some important points for enhancing residents' preparedness. First, we should consider the relationship between flood-risk perception and sense of self-responsibility. The sense of self-responsibility is important because it corresponds with residents' intention of preparedness against floods. Although flood-related learning activities in general have tried to enhance residents' flood-risk perception, giving information about why residents have to take responsibility for preparedness is not enough. To give a currency to people's responsibility as residents, the Pafrics developed by the NIED project (refer to Part III) is equipped with some information about residents' responsibility on flood preparedness. One of the basic functions installed in the Pafrics is based on the analysed models in this study. This is a newly constructed flood learning system because it is based on some social psychological analysis. Second, we should not miss the relationship between the fear of floods and the concern with floods. A determinant of the concern with flood is important because it relates directly to residents' preparedness. Although traditional flood learning has tried to enhance residents' feeling of fear, even if fear of flood is enhanced, it is not enough to motivate residents to prepare against floods. Actually, if residents are to prepare against floods, they need to know flood mechanisms. If residents were to take part in workshops that used PAFRICS, they might get an opportunity to reconsider resident-based flood prevention. Many scientists have emphasized the importance of resident-based disaster preventive plans, and our new education system will take an important step in that direction by providing residents opportunities to participate in such resident-based flood risk management.

References

Aichi Prefectural Government (2000): Prevention floods plans of Aichi Prefecture for fiscal year of 2000. Aichi: Aichi Prefectural Government (in Japanese).

Burton, I., Kate, R. W., and White, G. F. (1993): *The Environmental as Hazard* (2nd ed.). New York/London: The Guilford Press.

Cabinet Office (2002): *White Paper on Disaster Prevention.* Printing Bureau, Ministry of Finance, Japan (in Japanese).

Hashimoto, H., Matsunaga, K., and Nanri, Y. (2001): Flood disaster in Fukuoka City on June 29 1999. *Journal of Japanese Society for Natural Disaster Science*, **20**, 43–58 (in Japanese).

Lind, E. A. and Tyler, T. R. (1988): *The Social Psychology of Procedural Justice.* New York: Plenum.

Mizuno, H. (1981): Disasters and Information. Hirose, H. *A Social Scientific Approach to Disasters.* Tokyo: Shinyosha, pp. 83–109 (in Japanese).

Mulilis, J.-P., Duval, T. S. and Bovalino, K. (2000): Tornado preparedness of students, nonstudent renters, and nonstudent owners: issues of PrE theory. *Journal of Applied Social Psychology*, **30**, 1310–1329.

Takao, K. (2002): Perceived procedural fairness and the evaluation of authorities: A case of urban development. *Japanese Journal of Social Psychology*, **17**, 136–140 (in Japanese).

Takao, K., Maeda, M., and Nonami, H. (2003a): Residents' peception of procedural justice in implementing the rice terrace ownership system in Asuka Village, Nara Prefecture. *Journal of Rural Planning Association*, **22**, 26–36 (in Japanese).

Takao, K., Motoyoshi, T., Sato, T., and Fukuzono, T. (2003b): Do perceived flood risk and fear of floods directly affect preparedness?: The case of the Tokai flood disaster in Japan. The third annual international institute for Applied Systems Analysis-DPRI meeting on integrated disaster risk management: megacity vulnerability and resilience proceedings (http://idrm03.dpri.kyoto-u.ac.jp/Paperpdf/47takao.pdf).

Takao, K., Motoyoshi, T., Sato, T., Fukuzono, T., Seo, K., and Ikeda, S. (2004): Factors determining residents' preparedness for floods in modern megalopolises: the case of the Tokai flood disaster in Japan. *Journal of Risk Research*, **7**, 775–787.

Sjöberg, L. (1998): Worry and risk perception. *Risk Analysis*, **18**, 85–93.

Tyler, T. R. and Lind, E. A. (1992): A relational model of authority in groups. In M. Zanna (Ed.), *Advances in Experimental Social Psychology*, 25, (pp. 115–191). San Diego, CA: Academic Press.

Tyler, T. R., Boeckmann, R. T., Smith, H. J., and Huo, Y. J. (1997): *Social Justice in a Diverse Society*. Boulder, CO: Westview Press.

A better integrated management of disaster risks: Toward resilient society to emerging disaster risks in mega-cities, Eds., S. Ikeda, T. Fukuzono, and T. Sato, pp. 153–163.

Roles of Volunteers in Disaster Prevention: Implications of questionnaire and interview surveys

Isamu Suzuki

1 Introduction

In Japan, a country prone to natural disasters, community organizations such as volunteer fire corps and flood-fighting organizations have assumed a role in preventing disasters. To enable rapid early relief activities in the event of a disaster, local disaster prevention efforts must be made in normal times. Local residents as well as government authorities also need to participate and cooperate in devising ways to protect communities so that residents' efforts can be reflected in administrative planning. It has been pointed out (Cabinet Office, 2002), however, that the ability of communities to prevent disasters has declined as nuclear families have become predominant, traditional communities have declined, and the number of solitary elderly people has increased. Despite such decline of communities' preparedness to disasters, the number of people volunteering to participate in relief activities has increased in the event of disasters in recent years.

More than ten years have passed since the Hanshin-Awaji Earthquake of 1995, which was said to be the first year of volunteer activities in Japan, and volunteer activities in times of disaster are no longer matters worthy of special mention. Today, there are several nationwide networks established by disaster volunteer groups, and their relief activities performed through nationwide networking are proving successful. It is therefore important to study the activities that are actually performed by these disaster volunteers, the problems being encountered by them, and the role that they now play in disaster management in terms of facilitating communities' preparedness to disasters through all cycle of them.

To study these subjects, questionnaire surveys and interview surveys have been conducted. The questionnaire surveys were done in the areas affected by the Niigata Heavy Rain, the Fukui Heavy Rain, and Typhoon No. 23 and Heavy Rain in 2004. The interview surveys were done in the areas affected by the Niigata Heavy Rain and the Niigata-ken Chuetsu Earthquake in 2004.

In this chapter, first we summarize the history and state of disaster volunteers in Japan (Section 2). Second, on the basis of the results of the questionnaire surveys and interview surveys, we discuss current disaster volunteer activities and the challenges being encountered (Section 3). We then discuss what disaster prevention efforts should be like in the coming years on the basis of the study results obtained thus far (Section 4).

2 Past and Present of Disaster Volunteer Activities in Japan

Numerous volunteer workers have been involved in disaster relief since the Great Hanshin-Awaji earthquake of 1995 and the oil spill off the coast of the Japan Sea of 1997. Today, volunteers are essential to disaster relief. Nationwide networks of disaster volunteer groups have been established. The "National Earthquake Disaster Network" was founded in November 1997 composed of a dozen member groups nationwide. "Japan Disaster Relief Network (J-net)" was organized in January 2000 with twenty-odd members all over Japan (Fig. 1).

Disaster relief activities by national networks, although still at a trial-and-error stage, have been steadily producing effective results. For example, volunteers began exchanging information and helped establish volunteer centers in affected areas immediately after the Tokai Heavy Rain of 2000, the eruption of Mt. Usu in Hokkaido in 2000 and the earthquake centered in the northern part of Miyagi Prefecture in 2003. When floods occurred in many parts of the country in 2004, disaster volunteer centers were established in affected areas and volunteers participated in relief activities. According to published figures (FDMA, 2004; Ibaraki Social Welfare Council, 2004), the number of volunteers who participated in relief activities was about 45,000 in Niigata Prefecture at the time of the Niigata Heavy Rain, about 58,000 in the Fukui Prefecture at the time of the Fukui Heavy Rain, and about 22,000 in the Hyogo Prefecture at the time of Typhoon No. 23, respectively. From a practical viewpoint, it is of great significance that various groups have been seeking relief activities more focused on victims and making efforts to participate in timely activities via networks.

Behind the active formation of networks of disaster volunteer groups was the difficulty in maintaining volunteer activities under normal conditions. Boy scouts, YMCA and other numerous existing volunteer organizations were involved in relief activities different from those under normal conditions during the Great Hanshin-Awaji Earthquake (Nishinomiya Volunteer Network, 1995). In business circles, the labor union of Matsushita Electric Industrial Co., Ltd., COOP Kobe, and other groups delivered goods, prepared hot meals

Fig. 1. Locations of groups constituting J-net.

or were engaged in cleaning (Kaneya, 1996). Emergent organizations were created as networks of these groups. Typical examples include local NGO support coordination councils and "Nishinomiya volunteer network". Several other emergent organizations were created. Not many emergent organizations have, however, been continuing activities after the earthquake. Volunteer workers also participated in relief on such occasions as the Great Kanto Earthquake of 1923, Fukui Earthquake of 1948, Typhoon Ise Bay of 1959 (Noda, 1995; Urban Disaster Research Institute, Disaster Prevention Bureau, National Land Agency, 1987), and the pyroclastic flow disaster due to the eruption of Mt. Fugen, Nagasaki Prefecture in 1991 (Kanegae, 1993; Yamashita, 1994). Not many of such activities have been continued at normal times. As Dynes and Quarantelli (1968) pointed out some time ago, volunteers who work in emergent organizations return to their normal life once the urgent relief period is over. Extending organizations resume their daily activities. Then, these types of organizations have difficulty in continuing relief activities. To compensate for such drawbacks of volunteer work, disaster volunteer groups develop a national network to continue activities under normal conditions through coordination with other groups.

Excessively organized emergent groups through networking to continue activities is likely to reduce flexibility. In emergency conditions, including disasters, uncertainty increases (Noda, 1997) and one event occurs after another that is unexpected based on pre-prepared manuals or scenarios. What is most important in the field in disaster relief is therefore relief measures taken according to circumstances. For example, when distributing relief goods at a specific shelter or when providing relief services to specific elderly or physically impaired people, quick measures should be taken focused on immediate needs while no explicit specifications are available in manuals. Actions are required to modify available principles described in manuals when they are found to be inappropriate under actual conditions, or to work out response to events not assumed in manuals (Tachiki, 1999). Excessive organization of volunteers is likely to hamper such actions. Volunteer workers are expected to prove their potential in situations not assumed in disaster manuals or scenarios if they are not excessively organized but are allowed to work freely (Atsumi, 1998).

National networks of disaster volunteer groups therefore need to continue daily activities under normal conditions while keeping themselves prepared to take impromptu measures during a disaster under a situation not defined in predetermined relief programs. It is necessary to organize volunteers to enable them to take actions according to the circumstance during a disaster and continue activities under normal conditions.

Subsequent sections will focus on the above issue and discuss activities of volunteers during a disaster.

3 Disaster Volunteer Activities and Problems of Disaster Prevention

3.1 Purpose and method

A questionnaire survey and interview surveys were conducted to collect information on volunteer activities in times of floods or earthquakes. The questionnaire survey covered the cities of Sanjo (Niigata Prefecture), Fukui (Fukui Prefecture), and Toyooka (Hyogo Prefecture) which respectively suffered flood damage caused by the Niigata Heavy Rain, the Fukui Heavy Rain, and Typhoon No. 23 in 2004. From the Basic Resident Register, 1,000 households were chosen from each city by random sampling. Replies were obtained from 1,259 households (the response ratio was 42%).

An interview survey was conducted in September, 2004, in flood-affected areas in the Niigata Prefecture. The interviewees were members of the Sanjo City Hall, Sanjo City Social Welfare Council, Mitsuke City Hall, Mitsuke City Social Welfare Council, and the Nakanoshima Social Welfare Council. In February, 2005, an interview survey concerning the Niigata-ken Chuetsu Earthquake was conducted. The interviewees were members of the Nagaoka City Social Welfare Council, the former Yamakoshi Village Volunteer Center, and KOBE-kara Ouen-suru Kai (Supporters from KOBE; a disaster-relief NPO).

3.2 Results and discussion

According to questionnaire results, "mud removal" and "disaster debris removal" were the most common activities carried out by volunteers in the aftermath of floods, and both of these activities accounted for about 30% of all volunteer activities, followed by "assistance to evacuees" (13%). With respect to the questions concerning the respondents' opinions about volunteers, 72% of the respondents chose "Strongly agree" concerning the statement "Assistance provided by volunteers contributed to the recovery of the affected areas." Concerning the statement "I am grateful to the volunteers", 74% of the respondents chose "Strongly agree." These results indicate that the people in the affected areas have positive opinions of volunteers. The high percentages of respondents who felt grateful to the volunteers and who thought the volunteers contributed to the recovery of the affected areas indicate that the volunteers were not regarded as mere "feel-gooders", but as people who really helped the affected areas.

Disaster volunteer activities in Japan gathered momentum at the time of the Hanshin-Awaji Earthquake. During the 10 years that followed, a number of disasters occurred, and today volunteers participating in disaster relief activities are no longer a rare sight. In times of disaster, disaster-relief NPOs, government agencies, and local organizations in the affected areas, often jointly establish disaster volunteer centers. At the time of the recent Niigata Heavy Rain, five volunteer centers were established during the period from July 14 to 16 under the leadership of the social welfare council.

At the time of the Niigata flood, however, cooperation between volunteers and the administrative authorities posed a number of problems. The interview survey revealed that in areas which had not previously experienced disasters, cooperation between the government, disaster-relief NPOs, and local organizations such as social welfare councils was essential to carry out relief activities. Some confusion arose, therefore, when the disaster volunteer centers were established.

A major factor was the lack of know-how regarding cooperation between volunteers and local government. Because the use of volunteers in times of disaster was a novel step, each action was a new experience and took time to implement. Another factor that contributed to confusion was that the government and the social welfare council accepted volunteers separately.

In contrast, at the time of the Niigata-ken Chuetsu Earthquake, disaster volunteer centers were quickly established. During the period from October 24 to 30, 11 disaster volunteer centers were established in Niigata Prefecture under the leadership of the social welfare council. Websites were also opened, and information concerning volunteers and supplies was disseminated through the Internet. The ability to establish quickly the volunteer centers was related to the experience gained from the past flood disaster. First, a system to facilitate cooperation between disaster-relief NPOs from other areas, the government, and the social welfare council (which played an important role in establishing and operating the volunteer centers) was already established. Second, having learned from the past flood disaster, the government clearly identified the contact organization responsible for accepting volunteers.

The interview survey, however, revealed some remaining challenges related to volunteer activities. First, it was difficult to take appropriate and prompt action in a situation where no operational manual was available and unexpected events occurred one after another. To overcome this problem, appropriate coordination of personnel, goods and equipment, money, and information is needed. A second challenge is to find ways to make effective use of the knowledge accumulated in connection with the recent earthquake in future disasters within the same areas and elsewhere. For example, from experience acquired through this flood and earthquake, NPOs and social welfare councils in other prefectures have gained the know-how needed to smoothly accept volunteers. Thus, the challenge is to enable effective dissemination of such know-how so that it can be applied in future disasters regardless of where they occur. A third challenge concerns the need to train volunteer leaders who can work over an extended period of time. Volunteers arrive at an affected area determined to work hard to help, but there are not always suitable activities for them to participate in so they often remain idle for a long time. Volunteer leaders need to be trained so that they can cope with such situations. A fourth challenge concerns the need to turn over volunteer activities to local organizations or local NPOs. Sooner or later, volunteers have to leave the affected area during the emergency period. To provide long-term assistance and ensure recovery of the affected area, local organizations must be prepared to take over the volunteer activities.

3.3 Implications to the activities of disaster volunteers

Based on the above survey results, this section summarizes the main challenges regarding disaster volunteer activities and proposes measures to meet those challenges.

(1) Need for rear-echelon support

In times of disaster, information must be shared concerning personnel, goods and equipment, and funding, and emergency relief supplies and volunteers must be efficiently coordinated. At disaster volunteer centers established in an affected area, it is difficult to take the entire affected area into consideration because of the need to cope with changing conditions in the immediate vicinity. There is a need for rear-echelon organizations or groups that can liaise between disaster volunteer centers in different areas and coordinate the traffic of personnel, goods and equipment, and money to enable efficient relief activities.

(2) Need for long-term support

Evacuees may need to live in temporary housing for two or more years. At the time of the Hanshin-Awaji Earthquake, the deaths of elderly people living alone in temporary housing was reported later by mass media. To prevent similar incidents, long-term support must be provided to the residents of such temporary housing. Fortunately, there are many areas, such as in Nagaoka City and Yamakoshi Village, where disaster volunteer centers have been established within temporary housing sites. Such disaster volunteer centers need to provide long-term livelihood support to evacuees living in temporary housing in cooperation with the government, social welfare councils, social welfare organizations, local NPOs, and other organizations.

(3) Importance of the role of volunteers or NPOs in reconnecting the affected area with the rest of the world

In the interview survey, volunteers and NPOs were often likened to "adhesives" or "catalysts". That is, volunteers and NPOs were described as adhesives for connecting people together or connecting various activities together and as catalysts for reconnecting the affected area with the rest of the world. As various people and activities became connected, new relationships came into being. Relationships thus formed while dealing with a disaster together may have commercial and cultural significance and might help the affected area recover.

(4) Need for regional cooperation in sharing and accumulating disaster experience

The know-how gained from the experience of the Niigata-ken Chuetsu Earthquake is invaluable, and the knowledge thus gained must be used effectively to cope with future disasters. The acquired disaster experience and knowledge should be shared with the world, rather than simply retained in the affected area. To enable such sharing, it will be necessary to create regional cooperative relationships with outside organizations and groups.

(5) Need for cooperation in normal times

Disaster prevention is synonymous with community building. It is not a task that can be fulfilled by disaster-related organizations or groups working alone in times of disaster. For example, providing assistance to elderly people and people with disabilities requires cooperation between the various people who are engaged in welfare activities in normal times. Likewise, the rapid and smooth dissemination of information through the Internet in times of disaster requires the prior cooperation in normal times of people who are well versed in information technology and who have the knowledge needed to transmit relevant information. By building cooperative relationships among various local resources in normal times, activities in times of disaster can be carried out more smoothly and long-term assistance can be provided to affected areas.

4 Concluding Remarks: Ways to Enhance the Ability of Communities to Prevent Disasters

The surveys have led to the following major findings. Concerning problems and challenges related to volunteers in times of disaster, there is a need for rear-echelon support in times of disaster and long-term assistance in affected areas. Volunteers and NPOs play an important role of "catalysts". In addition, regional cooperation in sharing and accumulating knowledge gained from disaster experiences and prior cooperation in normal times are beneficial.

These findings make it possible to define an alternative approach to disaster prevention for the coming years. As mentioned, it is generally acknowledged that the ability of communities in Japan to prevent disasters has declined, as the decreasing number of volunteer fire corps indicates. However, local communities have to continue to play an important role in relief activities in times of disaster and in the subsequent recovery efforts. Thus, it is necessary to enhance the ability of communities to prevent disasters through approaches that differ from the conventional ones.

Important factors that help enhance the ability of communities to prevent disaster are acquired knowledge, information, and mutual trust. In a traditional territorial society, where almost all people are mutual acquaintances, the history of local disasters and the experiences of those disasters are handed down as wisdom, and information regarding dangerous sites and persons who need help in times of disaster is shared. In modern society, though, people's lives are isolated by their lifestyles. Newcomers to an area do not know the history of local disasters and, not knowing who lives where, they are unable to build relationships of trust. If alternative ties are to be created under such circumstances, there is a need for a new type of cooperation that enables people who are engaged in different types of activities to build ties with one another.

Disaster volunteer organizations can serve an important purpose in this situation. The history of disaster volunteering in Japan is relatively short, and disaster volunteer organizations are still sometimes "outsiders" in the eyes of traditional local organizations. Outsiders and mavericks, however, can cause major changes in existing communities (e.g., Sugiman, 2000). This is one way volunteers can serve as catalysts. It is possible, therefore, that disaster volunteer organizations will change existing ties in a local community so as to improve local disaster prevention activities. Local organizations (such as neighborhood community associations, volunteer fire corps, and volunteer disaster-prevention organizations), various NPOs (such as disaster-relief NPOs and social welfare NPOs), and local government authorities need to build a network through interorganizational cooperation (Fig. 2).

To enhance the ability of communities to prevent disasters, these organizations and individuals should share knowledge and information while also building mutual trust on the basis of the concept of risk communication. To do so, it is necessary to deal with various types of risk (e.g., risk of crimes, environment, welfare, and so on) instead of focusing only on disaster prevention. It is also worth keeping in mind that unilaterally conveying knowledge about flood prediction, etc., is not enough; local residents must also discover various local risks by themselves. To do this, it may be helpful for a disaster volunteer organization to hold, for example, a "town walking workshop" to learn more about the local community (e.g., Watanabe, 2000). A town walking workshop is an activity to identify dangerous sites and problems that may be encountered in the event of a disaster while walking in the local community. Activities like this make it possible to discover that there are various interrelated risks, including disaster risks, in the local community and to learn that these various risks must be addressed through various types of knowledge and activities.

As a result of exchanges between people working in different fields, people who have never met should be able to work together outside of the frame-

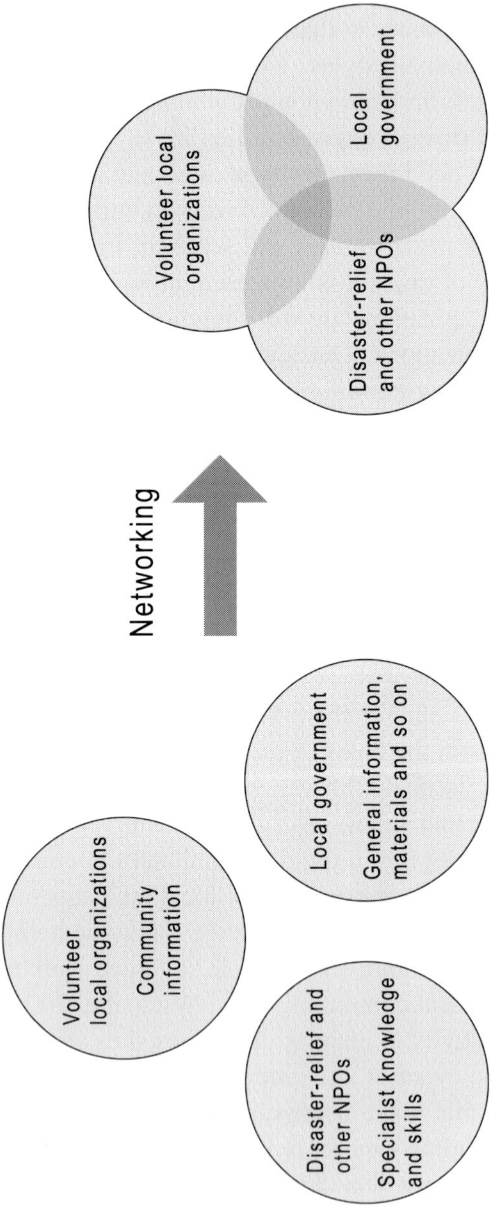

Fig. 2. Conceptual view of local network.

work of organizations and activities and develop ties that will help re-energize communities. Revitalization of communities will strengthen their ability to prevent disasters. The consequence will be to enable prompt and flexible responses to disasters and long-term assistance to victims through local disaster prevention efforts to protect communities from river flooding.

References

Atsumi, T. (1998): Future of disaster relief systems and volunteer activities, *Urban Politics* (Toshiseisaku), **92** (in Japanese).

Cabinet Office (2002): *2002 White Paper on Disaster Management*. Printing Bureau, Ministry of Finance (in Japanese).

Dynes, R. R. and Quarantelli, E. L. (1968): Group behavior under stress: A required convergence of organizational and collective behavior perspectives, *Sociology and Social Research*, **52**, 416–429.

Fire and Disaster Management Agency (2004): Volunteer activities related to the Niigata and Fukui heavy rain disasters (in Japanese) [on-line]. Available from http://www.fdma.go.jp/html/volunteer/naiyo.html.

Ibaraki Prefectural Social Welfare Council (2004): Disaster relief volunteer information related to Typhoon No. 23 and the Niigata-ken Chuetsu Earthquake, Vol. 4 (in Japanese).

Kanegae, K. (1993): *Roaring Fugen*, Shueisha Inc. (in Japanese).

Kaneya, N. (1996): Volunteers take the initiative—Volunteers in disaster-affected areas—in Honma, M. and Deguchi, M. (ed.) *Volunteer Revolution—From experience in great earthquake to civic movements*, Toyokeizai Inc., 19–42 (in Japanese).

Nishinomiya Volunteer Network (1995): How volunteers took activities at the Hanshin-Awaji Earthquake, Japan Broadcast Publishing Co., Ltd. (in Japanese).

Noda, M. (1995): *Disaster Relief*, Iwanami Shoten, Publishers (in Japanese).

Noda, T. (1997): *Disasters and Social Systems*, Kouseisha-kouseikaku Corporation (in Japanese).

Sugiman, T (2000): Aiming for an autonomous social system (in Japanese). In: Sugiman, T., editor. *Field Work Human Science: Reviving Communities*. Minerva Publishing, pp. 29–148.

Tachiki, S. (1999): Risk management by network organizations in Kuroda, N., Tsuganezawa, T. (ed.), *Kanseigakuin University Great Hanshin-Awaji Earthquake series II: Sociology of earthquake disasters—Great Hanshi-Awaji Earthquake and public awareness*, Sekaishisosha Co., Ltd., 135–158 (in Japanese).

Turner, B. A. (1978): *Man-made Disasters*, Wykham publication.

Urban Disaster Research Institute, Disaster Prevention Bureau, National Land Agency (1987): Study report on disaster volunteer activities (in Japanese).

Watanabe, T (2000): A fifth-year experiment in local community: a study on the practice of "local disaster prevention that's not called local disaster prevention" and group dynamics (in Japanese). *Japanese Journal of Experimental Social Psychology*, **39**(2), 188–195.

Yamashita, Y. (1994): Long-term disasters and local community—Case study of Shimabara City, Nagasaki Prefecture -, Social Analysis, **21**, 109–123 (in Japanese).

A better integrated management of disaster risks: Toward resilient society to emerging disaster risks in mega-cities, Eds., S. Ikeda, T. Fukuzono, and T. Sato, pp. 165–176.

Issues and Attitudes of Local Government Officials for Flood Risk Management

Kiyomine Terumoto

1 Introduction

Flood risk management is currently in transition and there are several issues involved: 1) Not only flood control structures but also non-structure measures in flood risk management have been considered as being important. 2) A shift from the concept of building flood control structures in order to prevent flood disasters to the direction that allows a certain flood risks is emerging. 3) There is a social situation in which the necessity to offer risk information including flood risk, responses based on self-responsibility, and the importance of citizen participation have been recognized.

The objectives of this chapter are to analyze the awareness related to the current flood risk management issues among local government officials, based on the survey. Then, issues and attitudes of local government officials for promotion of flood risk management in the future will be examined.

2 Reviews of Current Issues with Regard to Flood Risk Management

2.1 Provision of public funds to disaster victims

The Natural Disaster Victims Relief Law, which was enforced in November 1998 to provide money for disaster victims in order to support their livelihood, was revised in April 2004. Following the revision of the law, a housing stabilization support system was implemented to assist with the expenses related to housing repairs and land leveling. However, the law does not cover the actual costs of reconstructing the main bodies of buildings. Also, there are income and age restrictions in the law. In addition, regarding the recognition of the damage, support is not provided to the victims of inundation above floor level, unless the building damage is recognized as "complete collapsed" or "large-scale half-collapsed".

What the damaged areas want does not agree with what the support systems based on the law provide for them. In addition, support depends on how the local government recognizes the degree of the damage. Providing money directly to the victims is actually effective in advancing the reconstruction of the damaged area. On the other hand, generous support for the victims may create moral hazards, and it may discourage the promotion of disaster reduction measures. Well-balanced systems in which support is fairly provided and the needs of the victims are taken into account are necessary.

2.2 Development of the awareness of self-responsibility among citizens

When a large-scale disaster occurs, it is impossible for administrative officials to carry out disaster-related responses only by themselves. Disasters have to be reduced through self-help by each citizen and cooperation by citizens. However, many citizens don't take disaster-reduction action, even under crisis and depend on the government to do that in Japan. So, in terms of disaster-reduction measures, citizens' awareness of their self-responsibility is a big problem. Therefore, disaster education is an important factor in increasing disaster-reduction awareness among citizens. However, it is also pointed out that local government officials tend to see the ineffectiveness of ordinary disaster education and material distributed to citizens (Terumoto *et al.*, 2004). Regarding disaster education, disaster-reduction measures and the roles of citizens in disaster-risk mitigations must be clarified.

2.3 Expansion of authorities of the local governments regarding river improvement

As rivers form a network, consistent management of river systems is effective. On the other hand, rivers have important functions in local areas, such as amenity facilities and landmarks. So rivers in local areas should be regarded as a part of land-use planning.

Local government authority over river improvement was expanded in 2000. As a result, local governments were able to carry out certain levels of river improvement to the first-grade rivers that are under their direct jurisdiction. However, items such as the establishment of primary river improvement plans are not included in the local government authorities. River improvement management should be linked with local areas and with each management unit, in terms of various functions related to flood control, irrigation, and the environment.

2.4 Publishing of flood hazard map

Publishing risk information on areas which might have high flood risk is requested from the social point of view. For instance, citizens should have the right to know, and should be able to choose some options for the disaster measures based on the information. Such information is also expected to raise disaster reduction awareness and increase citizens' knowledge.

Publishing risk information of high flood risk areas in designated rivers has been the obligation of local governments, through the revision of the Flood Prevention Law in 2001. Furthermore, based on the fact that many flood disasters occurred in 2004, the Flood Prevention Law was revised again in 2005, expanding the range of designated rivers, which included small-to-medium-sized rivers. However, only 434 local governments had made and published flood hazard maps as of September 2005 (MLIT, 2005).

On the other hand, hazard maps have various problems. For example, 1) it is extremely difficult to identify property flood risks, 2) hazard maps reflect only some of the extreme cases, 3) ordinary hazard maps do not take into account internal flooding, 4) it is worrying that inaccurate information may affect the value of real estate, and, 5) there are other risks that the information on hazard maps may be treated as "safety information". So, there are various barriers against using hazard maps effectively.

2.5 The disclosure of flood risk information on real estate transaction

Land which had been originally flood control basins is targeted as new development areas, because the number of designated developmental areas have decreased due to urbanization. Consequently, in some cases new development areas have high flood risks. Many citizens may choose such areas as their new places to live without knowing about flood risks and other characteristics. They want to know and should know how high the risk of flooding is in their area. Therefore, it is important that the sellers are forced to disclose any relevant hazard information to the buyers at the time of real-estate transaction.

Under the Real Estate Business Law in Japan, there are items related to the disclosure of important information at the time of real-estate transactions. A question of whether the land is in landslide disaster warning areas is also included among these items. However, other natural disasters are not included. The disclosure of the information on other natural disasters should be also discussed.

2.6 Land-use restrictions in flood risk mitigation measures

Damage caused through internal flooding is increasing in urban areas in Japan. This is caused by the fact that the ability to drain rainwater has decreased in river basins due to urban development. Another reason is the fact that river improvement projects and regional development projects have been planned individually under vertically-segmented administrative systems. For this reason, such projects have been carried out individually without any mutual consistency.

New development areas or high flood risk areas should be actually targeted at land use restriction areas in flood risk mitigation measures. Currently, some local governments, although the number of such local governments is limited, are imposing building restrictions in high flood risk areas, based on the Building Standards Law in Japan. Regarding land-use management for flood disaster-reduction measures, the main issue is to appropriately control land-use to keep pervious areas from unregulated development. It is necessary to create such a scheme to manage urban and regional planning linked with the flood control measures.

2.7 Necessity to make plans based on citizen participation

It has become more and more necessary for citizens to participate in procedures in making plans regarding not only river improvement projects but also overall public projects. Major river improvement projects must reflect residents' opinions according to the River Law. Based on these social and systematic requests, river improvement projects reflecting citizens' opinions have been carried out in various areas. On the other hand, there are still more cases in which citizens only partially participate in the processes. In other words, in such cases project plans are completed in advance, and those plans will be hardly changed, even though residents receive explanations on the project at a meeting afterwards.

River improvement project with citizen participation is currently at its germinal stage in Japan. The accountability to citizens of related various projects and the transparency of planning procedures have become increasingly needed, and citizen participation has also become important. There is another aspect that river improvement planning requires high expertise. So decisions cannot always be made in terms of reflecting the ideas and opinions of citizens. It is necessary to match the citizens' ideas and opinions and experts' special knowledge.

2.8 Improvement of interface functions for planning support

In order to offer a common platform where citizens, administrative officials, and experts could cooperate to draw up the plans for flood disaster-reduction measures, some supportive systems are required. For meaningful discussions, residents' ideas and opinions, based on reasonable understanding of flood risks in local areas, cost-effectiveness regarding flood disaster reduction measures, and other important factors, are required.

Recently, many government authorities have been putting out information about flood hazards on websites, as the Internet is far-reaching. Some government offices are carrying out trials of holding discussions related to the project plans via online forums. Additionally, the provision of information regarding flood risks through the use of GIS is an effective function. There is a report that shows the information provided on disaster risks through GIS changed citizens' awareness of disaster risks (Kawasaki *et al.*, 2003).

Flood risks and cost-benefit of disaster reduction measures are generally unclear to citizens. So it is effective to offer the information by using various tools and to exchange opinions based on that information, in order not only to improve citizens' awareness of flood disasters, but also examine flood disaster-reduction plans that reflect citizens' ideas and opinions. These efforts are still at the infant stage, but similar efforts will be made in various areas in the future.

3 Methods and Materials

The survey was carried out among officials in river management and emergency response sections of local governments in Japan. (There were 3,238 local governments at the time of the survey, including the 23 wards of Tokyo.) Questionnaires were mailed to each local government on February 13 and 14, 2003, and they were to be returned by mail by March 24, 2003.

The official who has worked the longest in the section was requested to answer, regardless of his/her job class. In those instances where one section was in charge of both river management and emergency responses, each different official at the section took care of answering the questionnaires. The response rate was 40.2% among river management sections and 46.6% among emergency response sections.

In the awareness of flood risk management, it was anticipated that there would be differences between local governments with high flood risks and those with relatively low risks. Accordingly, the differences of the awareness based on their disaster experience in each local government were also analyzed. As to the records of flood disasters, the number of disasters (which had affected each local government for the last five years before the survey) was figured out, based on flood damage statistics (MLIT 2000–2004). In the analyses, the number of disasters was classified into three categories: zero disaster, one disaster, and two or more disasters. The attributes of respondants in terms of their age and disaster experience are shown in Figs. 1-1 and 1-2.

Eight questionnaire items related to flood risk management, which were reviewed in Section 2, were investigated in the survey. The items are as follows.

i) To spend taxpayers' money to compensate damages for disaster victims.

ii) To restrict building structures and uses in high flood risk areas to reduce flood risks.

iii) To develop citizens' awareness of self-responsibility.

iv) To impose disclosure about flood risk information for sellers or lenders of properties, relating to real-estate transactions in high flood risk areas.

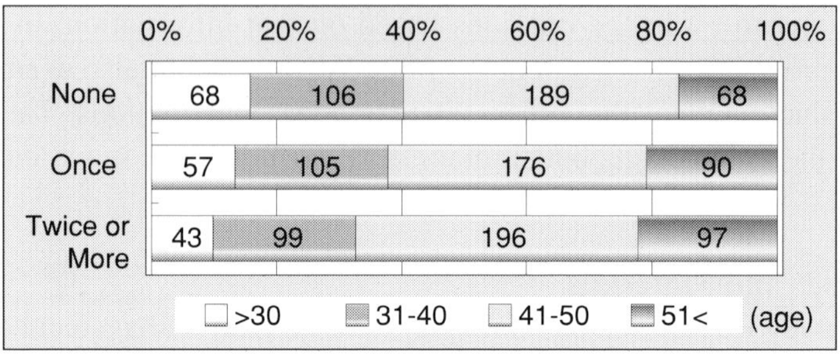

Fig. 1-1. Attribute of Respondents in terms of age and flood disasters experience (Officials in River Management section).

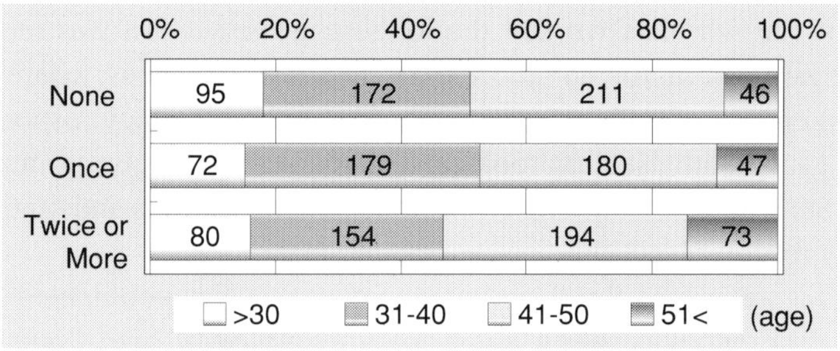

Fig. 1-2. Attribute of Respondents in terms of age and flood disasters experience (Officials in Emergency Response Section).

v) To publish risk information on hazard maps or other materials.

vi) To expand the local government authority for river improvement.

vii) To make river improvement project plans through citizen participation.

viii) To create interface functions to support planning activities for flood risk mitigation among citizens, administrative offices, and experts.

4 Results

4.1 The awareness of flood risk management

The responses were measured on a seven-point scale where "7" means "extremely necessary" and "1" means "not necessary at all", for each item. In the following analyses, seven points were assigned to the answer "extremely necessary", and one point to the answer "not necessary at all". The ordinal scale was treated as an interval scale.

Table 1. The awareness to questionnaire items in relation to flood risk management.

	River Management Section		Emergency Response Section		
Items	Mean	SD	Mean	SD	
To spend taxpayers money to compensate damages for disaster victims.	5.08	1.41	5.23	1.26	f=9.278**
To restrict building structures and uses in high flood risk areas to reduce flood risks.	4.88	1.32	4.85	1.19	f=0.438
To develop citizens awareness of self-responsibility.	5.66	1.11	5.87	1.09	f=26.03***
To impose disclosure about flood risk information for sellers or lenders of properties, relating to real-estate transactions in high flood risk areas.	5.28	1.35	5.27	1.27	f=0.002
To publish risk information on hazard maps or other materials.	5.41	1.23	5.41	1.18	f=0.002
To expand the authority of local government for river improvement.	4.71	1.17	4.70	1.11	f=0.061
To make river improvement project plans through citizen participation.	4.57	1.18	4.59	1.13	f=0.137
To create interface functions to support planning activities for flood risk mitigation among citizens, administrative offices, and experts.	4.67	1.18	4.74	1.10	f=2.961

***p<0.1% **p<1% *p<5%

The means and standard deviations of responses to each item are shown in Table 1. One-way analyses of variance for each item were conducted between the awareness of river management section officials and emergency response section officials (see Table 1). Moreover, to see the differences between the attributes, one-way analyses of variance (multiple comparison: Tukey's HSD test) were carried out in each section. The result of the analyses indicates a significant difference at the 5% level. The means and standard deviations and the results of one-way analyses for each attribute in each section are shown in Table 2.

Table 1 shows that item iii) marks the highest mean in both the river management sections and the emergency response sections. Especially among emergency response sections, the mean is relatively high. And also, Table 2 shows that the attribute of the "two or more disasters" is higher than "none". This result seems to be caused by local government officials' perception of the effectiveness in the experience of flood-disaster responses.

Item v) marked the second highest mean (see Table 1). Regarding item v), Table 2 shows that the local government officials in the attribute of the "two or more disasters", tend to be more aware of the necessity for the disclosure of high flood risk areas, compared to other local government officials.

Item iv) marked the third highest mean both among river management sections and emergency response sections. This result indicates that making

Table 2. Comparisons among disaster experience.

Items	River management section				Emergency response section			
	None	Once	Twice or more		None	Once	Twice or more	
	5.10	5.12	5.00	n.s.	5.29	5.26	5.14	n.s.
	4.78	4.82	5.03	None, Once<Twice or more	4.78	4.87	4.90	n.s.
	5.54	5.70	5.73	None<Twice or more	5.71	5.94	5.98	None, Once<Twice or more
	5.18	5.23	5.43	None<Twice or more	5.22	5.31	5.30	n.s.
	5.25	5.43	5.55	None<Twice or more	5.30	5.40	5.53	None<Twice or more
	4.61	4.69	4.83	None<Twice or more	4.63	4.76	4.72	n.s.
	4.56	4.57	4.59	n.s.	4.57	4.58	4.62	n.s.
	4.62	4.64	4.75	n.s.	4.66	4.75	4.83	None<Twice or more

it mandatory to disclose the information on high flood risk areas is required by local government officials. In addition, item ii), "building restrictions in high flood risk areas", shows a relatively low mean compared to item iv) ($t = -10.915$, $p < 0.001$ for river management sections; and $t = -13.82$, $p < 0.001$ for emergency response sections). Local government officials tend to be disinclined to impose building restrictions than to make it mandatory to disclose the information.

Item i) related to compensation for damage was also high means. The mean value was higher among emergency response section officials than river management section officials (see Table 1). As to item vi), local government officials in the attribute of the "two or more disasters" of river management sections tend to perceive it more necessity to expand local government authorities (see Table 2).

Compared to the other items, items vii) and viii) scored relatively low means. This result seems to show that citizen participation is less attention by local government officials.

4.2 Relations among the awareness of each flood disaster measures

In order to find the relevance between the items, correlation analysis was used. The correlation coefficients of the river management sections and emergency response sections are shown in Tables 3-1 and 3-2 which indicate that there is a strong correlation between items vii) and viii), items vi) and vii), items vi) and viii), items iv) and v), and items ii) and iv), respectively. The correlation coefficient between items vii) and viii) is the largest both among the river management sections and emergency response sections. These two items and item vi) are related to river improvement planning and citizen participation. The correlation between items iv) and v) is the relevance to disclose risk information. In addition, the correlation between items ii) and iv) is re-

Table 3-1. Correlation Coefficients (River Management Section).

	Items	i)	ii)	iii)	iv)	v)	vi)	vii)	viii)
i)	To spend taxpayers money to compensate damages for disaster victims.	1.00							
ii)	To restrict building structures and uses in high flood risk areas to reduce flood risks.	0.35	1.00						
iii)	To develop citizens awareness of self-responsibility.	0.20	0.40	1.00					
iv)	To impose disclosure about flood risk information for sellers or lenders of properties, relating to real-estate transactions in high flood risk areas.	0.21	0.52	0.46	1.00				
v)	To publish risk information on hazard maps or other materials.	0.19	0.42	0.46	0.58	1.00			
vi)	To expand the authority of local government for river improvement.	0.23	0.33	0.30	0.38	0.46	1.00		
vii)	To make river improvement project plans through citizen participation.	0.27	0.29	0.23	0.30	0.39	0.51	1.00	
viii)	To create interface functions to support planning activities for flood risk mitigation among citizens, administrative offices, and experts.	0.26	0.36	0.31	0.37	0.47	0.53	0.73	1.00

Table 3-2. Correlation Coefficients (Emergency Response Section).

	Items	i)	ii)	iii)	iv)	v)	vi)	vii)	viii)
i)	To spend taxpayers money to compensate damages for disaster victims.	1.00							
ii)	To restrict building structures and uses in high flood risk areas to reduce flood risks.	0.29	1.00						
iii)	To develop citizens awareness of self-responsibility.	0.25	0.35	1.00					
iv)	To impose disclosure about flood risk information for sellers or lenders of properties, relating to real-estate transactions in high flood risk areas.	0.20	0.52	0.44	1.00				
v)	To publish risk information on hazard maps or other materials.	0.24	0.41	0.43	0.50	1.00			
vi)	To expand the authority of local government for river improvement.	0.26	0.37	0.27	0.37	0.43	1.00		
vii)	To make river improvement project plans through citizen participation.	0.19	0.27	0.20	0.29	0.37	0.44	1.00	
viii)	To create interface functions to support planning activities for flood risk mitigation among citizens, administrative offices, and experts.	0.22	0.31	0.30	0.34	0.45	0.47	0.68	1.00

lated to the restrictions.

5 Discussion

5.1 Citizen participation in flood risk management

Developing citizens' self-responsibility is a crucial issue among local government officials. This tendency was especially strong among local governments stricken by flood disasters. Publishing risk information is also relatively crucial. In the current hazard maps, however, there are many problems to be resolved in this respect (Section 2.3).

Currently, new efforts, such as making new hazard maps through citizen participation and outreach activities, are being made in various areas in Japan.

When citizens made hazard maps by themselves as a part of flood disaster reduction measures, they would come to understand the nature of flood risks in each area. It is also effective to develop citizens' self-responsibility through these outreach activities. So, it becomes more necessary to provide the information regarding flood risks and other matters to citizens, as river improvement plans produced with citizen participation is going to be more frequent and be diversified. However, the requirement of citizen participation by local government officials was relatively low. It seems that the importance of risk communication between citizens and local government officials has not been well recognized so far by government officials.

5.2 Provision of public funds for disaster victims and the trend of local governments that suffered natural disasters

The result of the survey showed that compensation for victims is considered to be necessary by local government officials. This result is different from the basic concept of the national government's system that does not compensate private property.

On the other hand, regarding the conditions of the application of the Natural Disaster Victims Relief Law, for example, Tottori Prefecture instituted its own regulations after the Western Tottori Prefecture Earthquake and provided a maximum of 3 million yen for each household to cover its rebuilding costs. And other local governments have individually expanded support systems for victims after great disasters. This seems to be caused by the fact that the details of the restrictions under the law do not correspond with the needs of disaster-stricken local governments. The problem was seen in the awareness of local government officials. Arrangements are necessary to implement support measures under appropriate local conditions and disaster situations.

5.3 Importance of flood risk information

Currently in Japan, the disclosure of flood-risk information is not necessarily obligated at a time of real-estate transactions. The survey results, however, show that the necessity of obligation related to disclosure was perceived by the local government officials. Local government officials tend to think that the disclosure of information about high flood risk areas at the time of real estate transactions should be mandatory.

There are examples of this practice in foreign countries. In California, when a parcel of land or a building subject to high flood risk is traded, the seller must notify the buyer that the property is in special hazard area. If the buyer suffers damage to the property and the seller failed to provide enough information, the seller will be liable for the damages in some cases.

There remain many problems in making accurate hazard maps in terms

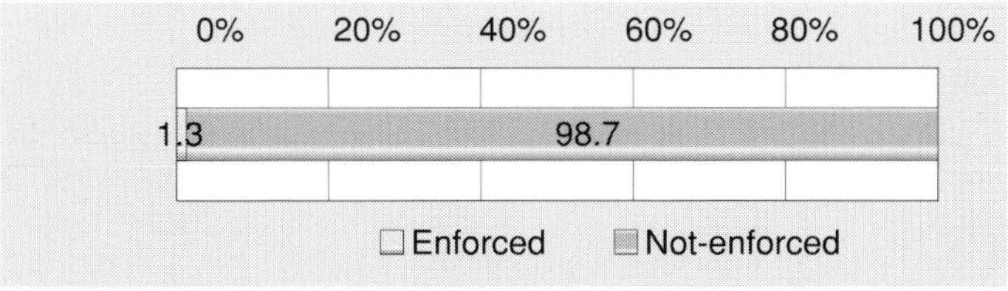

Fig. 2. Enforcement of Building Restrictions in Areas prone to be flooded.

of flood risk information, as indicated in Section 2.4. However, in order to enable buyers to receive risk information in Japan, making it mandatory to provide flood risk information at the time of a real-estate transaction must be examined.

5.4 Building restrictions in flood risk management

The response for "enforcement of building restrictions for flood risk mitigation in the areas prone to be flooded" is shown in Fig. 2. This result was based on another questionnaire item for officials in the river management sections. Approximately 1% of local governments have imposed building restrictions in areas prone to be flooded (see Fig. 2). Results indicate that an extremely small number of local governments have imposed building restrictions, while many local government officials considered it necessary to impose restrictions in areas prone to be flooded.

It seems really difficult to implement building restrictions which restrict public rights at the level of local government in Japan. Nevertheless, building restrictions for flood risk mitigation was considered necessary by local government officials who are to be responsible for protecting the lives and properties of citizens. If the necessity of building restrictions in local governments was recognized based on careful consideration, building restrictions should be examined with sufficient explanations of flood risk to residents.

6 Concluding Remarks

This chapter examined problems and prospects of flood risk management based on a survey and reviews. The results of the survey showed that there are still a number of gaps between the requirements of local government officials and current regulatory institutions in Japan. Bridging the gaps is a future task in the integrated flood risk management. For that purpose, clarifying the government authorities' and citizens' responsibilities is an essential problem.

References

http://www.mlit.go.jp/kisha/kisha05/05/051206/01.pdf (in Japanese).

Kawasaki, A., S. Yoshida, and S. Sadohara (2003): The influence of GIS based hazard map on people's attitude to natural disasters, *Journal of Architecture and Planning*, **569**, 109–115 (in Japanese).

Ministry of Land, Infrastructure and Transport; River Bureau (2000–2004): Flood Damage Statistics.

Terumoto, K., T. Sato, T. Fukuzono, and S. Ikeda (2004): Municipal employee's awareness of flood risks and flood mitigation measures, *Infrastructure Planning Review*, **21**, 335–340 (in Japanese).

A better integrated management of disaster risks: Toward resilient society to emerging disaster risks in mega-cities, Eds., S. Ikeda, T. Fukuzono, and T. Sato, pp. 177–192.

The Niigata Flood in 2004 as a Flood Risk of "Low Probability but High Consequence"

Teruko Sato, Teruki Fukuzono, and Saburo Ikeda

1 Introduction

On 13 July 2004, torrential rains fell over the Chuetsu region of the Niigata Prefecture that were remarkable for the total rainfall and amount of rainfall within a relatively short period of time that triggered hundreds of landslide events-collapsing hillsides, mudslides, and debris-and in the alluvial lowlands, the Ikarashi, the Kariyata, and other small to medium-sized rivers feeding into the Shinano River broke through their levees. Large-scale flooding devastated Sanjo City, Mitsuke City, and Nakanoshima Town where 16 people died, 2,500 hectares of land were inundated, 29 homes completely destroyed, 158 homes partly destroyed, 13,289 homes with major water damage, and 6,199 other buildings severely damaged. This disaster is referred to as the Niigata flood of 2004 or simple the flood.

The flood was primarily caused by the failure of the local levees, a highly unusual low probability but high consequences (LPHC) event because the levees are built to withstand torrential rain-induced flows. We begin by exploring the nature and exposure to flood hazard and by defining a damage and loss-risk structure framework. Then in the rest of the paper, we will describe the actual conditions and particulars of the Niigata flood risk.

2 Region and Drainage Environment of the Study

2.1 Topography

The Niigata Plain is an alluvial plain formed by centuries by flooding of the Shinano River. The plain is flat and narrow, sloping from north to south (see Fig. 1) with beach ridges and sand dunes to the west (i.e., on the Japan sea side) and hilly and higher mountains to the east. Starting at an elevation of about 19 m in the vicinity of Nagaoka City, the plain slopes gradually at a rate of about 0.7 to 0.8 m per 1,000 m and continues until the Nagaoka lowlands to the south and the Shinano River lowlands to the north (refer to the topography map). The Shinano River lowlands consists of formerly filled lagoons in which drainage is deficient and are frequently subject to flooding.

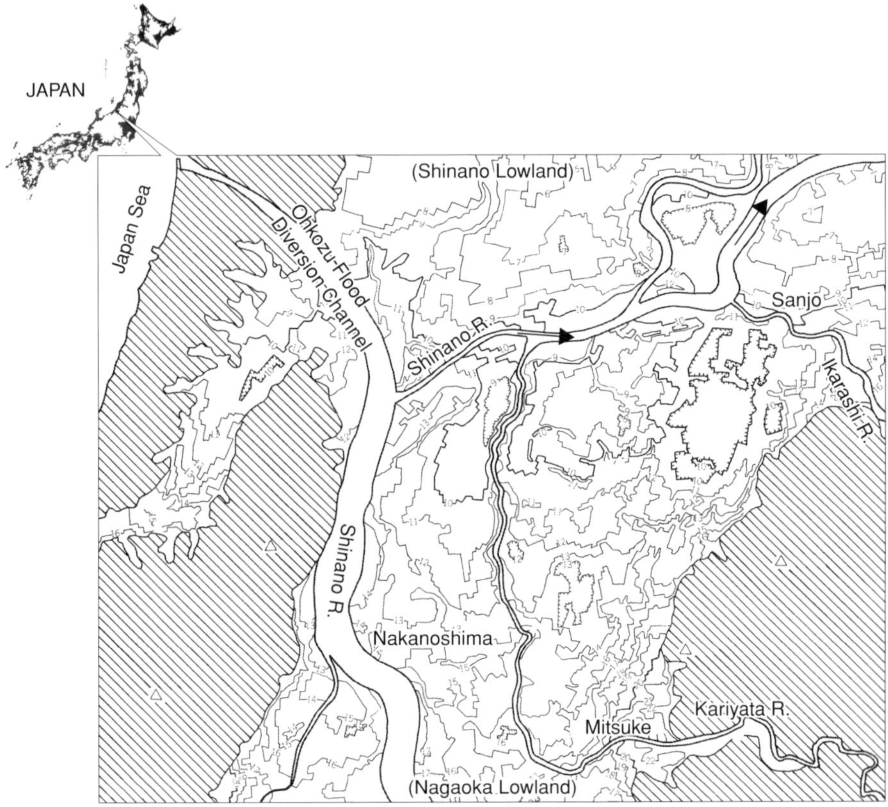

Fig. 1. Contour map of the alluvial plain of the lower reaches of the Ikarashi and Kariyata Rivers in the Chuetsu region.

The section of the Shinano River from the intake of the Okozu Flood Diversion Channel to the river mouth is called the lower reaches of the Shinano River. When the water level is low, the lower reaches of the Shinano River functions as the lowest downstream section of the Shinano River with a channel length of 367 km and a drainage area of 11,900 km^2. But when flooding occurs, all the flood waters flow from the Okozu Flood Diversion Channel to the Japan Sea (Fig. 1), and the lower section of the Shinano River becomes a major catchment basin that is independent of the upper reaches of the river. When this occurs, the drainage area shrinks to 1,420 km^2 and channel length to 58.2 km. In other words, under flooding conditions, the Kariyata and Ikarashi Rivers function as the uppermost tributaries feeding into the lower reaches of the Shinano, such that the upper section of the Shinano has little

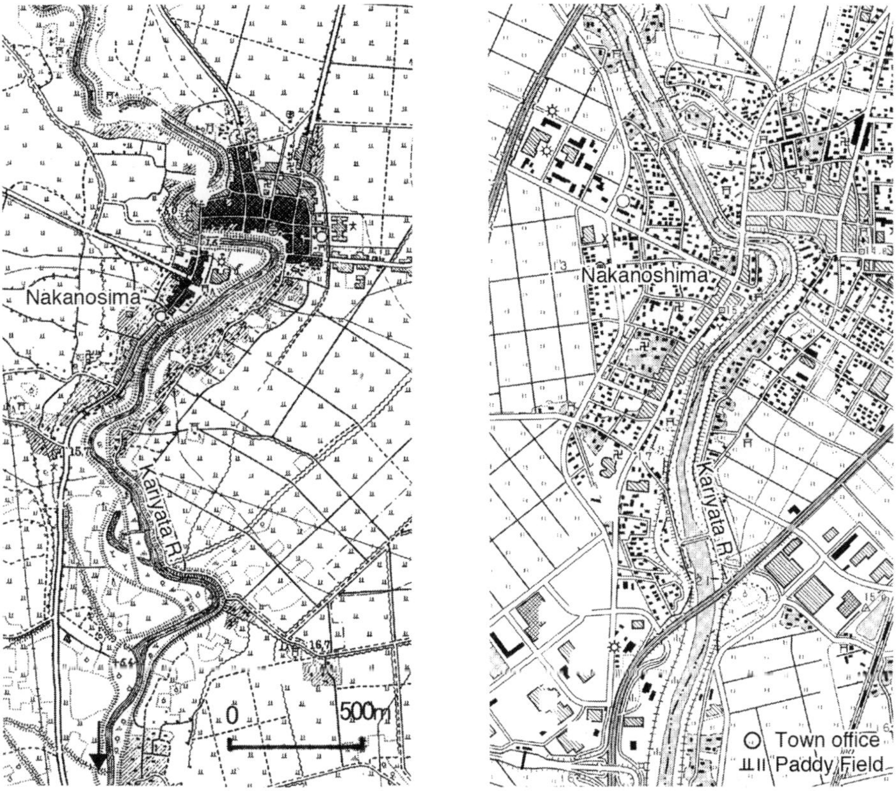

Fig. 2. Change of the land use of the Nakanoshima Town along the Kariyata River from 1948 to 2001.

effect on the flood runoff to the lower section of the river, while the two tributaries -the Ikarashi and the Kariyata- have an enormous impact on flooding in the lower drainage of the Shinano River. On the other hand, the slope of the lower reaches of the Shinano River and the two tributaries is gradual, so when the water level of the Shinano River rises, the backwater effect impedes the flow of the tributaries and the flood flow.

2.2 Flood history: social and land use changes and increased damage potential

Flood flows and the incidence of floods on the lower reaches of the Shinano River have markedly decreased since completion of the Okozu Flood Diversion Channel in 1931, but major flooding sometimes occurs on the Ikarashi and Kariyata tributaries. Levees failed due to flooding on both rivers in 1964, and levees along the Ikarashi River were breached in 1978. Thus, the alluvial lowlands of the lower reaches of both these rivers have seen floodins in the past and today, although flooding occurs less frequently, there is nevertheless a potential flood hazard in this area.

Figure 2 shows side-by-side topographical maps of Nakanoshima Town in 1948 and 2001. Comparing the two maps, one can see the extent that housing and factories and warehouses have encroached upon what were formerly wetlands used as rice paddies. It is also apparent that the Town office formerly built on a natural levee along the river at an elevation of 15.4 m has now been moved to reclaimed land on infilled wetland. The old commercial district of Sanjo City developed along the natural levee on the right bank of the Ikarashi River, but now the newer district of Rannan, has been built on the left bank and the commercial district pushed out into former paddy lands. In Mitsuke City too, there are ongoing efforts to upgrade and improve the river channel, the former now-abandoned river channel has been converted for commercial purposes, and paddy fields on the left bank of the river are also giving way to commercial development.

2.3 Awareness and responsiveness to flood risk: low probability but high consequences events

The 2004 flood was the first time in 40 years that the Chuetsu area has been by such massive flooding due to the failure of levees. Yet the potential for large-scale flooding is still clearly there. Given the region's potential for massive destruction from flooding, what is the degree of awareness of the local governments and inhabitants to their vulnerability to flooding?

Regarding flooding of the Ikarashi River, Sanjo City's "Community Disaster Prevention Plan 2003 Revised Version" states that "· · ·the Ikarashi River is not very safe. Although some flood control is provided by upstream dams, there are occasions when the levees leak or even fail as a result of heavy rains. Furthermore, when the water level of the Shinano River rises, this impedes drainage of the Ikarashi River into the Shinano River , which causes the level of the Ikarashi to rise and increases the danger of flooding." In addition, the "Sanjo City Flood Control Plan" notes that, although there are eight places along the Ikarashi River susceptible to flooding, critical flood controls are being implemented because levee failure is clearly a prospective danger. And yet in our interviews with local officials in Sanjo City, Mitsuke City, and Nakanoshima Town, we did not find one official who foresaw the failure of the levees. In addition, we noticed that all the recent disaster-prevention drills carried out by the City of Sanjo assumed an earthquake as the disaster, and no drills based on the assumption of a flood had been carried out for quite a long time.

The local inhabitants also felt secure in believing that the Ikarashi River would not flood, but, based on a fact-finding survey of people living in the stricken area gauging people's response to proposed river infrastructure improvements decided after the flood, their confidence had been expressed more symbolically. According to the Sanjo City Director of Public Works, "In 1996 when the previous Mayor was in office, the city in cooperation with the prefecture handed out a pamphlet detailing an Ikarashi River infrastructure improvement plan to every household in the district and warned people that the river was narrow and definitely posed a danger, but the response was pretty lukewarm. The fact that dams had been built upriver and no flooding had occurred for 40 years pulled most people into a false sense of security."

The National Research Institute for Earth Science and Disaster Prevention (NIED) conducted a questionnaire survey after the flood (Research Project on Societal Systems Resilient Against Natural Disasters, 2005), and asked people in the community "Do you know of anything that the volunteer Flood Brigade did to alleviate the disaster?" About 30% of the respondents said that they "did not know anything that the Flood Brigade did to help the situation," a percentage that increases to 50% if you include the respondents who "couldn't say one way or another whether the Flood Brigade helped". We also got a mixed response when we asked people if the "actions of the Flood Brigade were effective in mitigating hardship from the flood?" About half the respondents felt that the Flood Brigade was "somewhat effective" or "very effective" (42.2%), but an even larger percentage of people felt that the Brigade's actions were "not especially effective" or "can't say if their actions were effective or not".

Based on these responses, it is apparent the people's awareness about the risk of flooding and their concern about the activities of the Flood Brigade are fairly low.

3 Response to Flood Risk

3.1 Flood Mitigation Response by the Government

3.1.1 Tangible flood control initiatives

Table 1 shows the milestones of flood control infrastructure improvements on the Ikarashi River. Levee building on the Ikarashi was spurred by a major flood in August 1872, and construction of the levee on the left bank of the river was completed in 1877. A major river infrastructure improvement project was begun in November 1933 after another disaster in 1926 and by 1937, the river control infrastructure that you see on the Ikarashi today was more or less in place. Major floods were averted after that, but torrential rain and flooding in August 1961 resulted in the Mikura and Watarase bridges being washed out,

Table 1. Changes of the design flood discharge of a river channel.

Year	Design Flood Discharge (m^3/sec)	Design Flood Discharge of a river channel (m^3/sec)	Contents of Improvement works	Triger of the work
1875	-	-	Construction of levee on the left side of the river in lower reaches (completion in 1977)	Aug. 1872 Flood
1933	-	1,120	River improvement works, Construction of levee, and bank protection work (completion in 1926)	July 1926 Flood
1961	2,000	1,600	Construction of Kasahori dam (completion in 1964)	Aug. 1961 Flood
1980	3,600	2,400	Construction of Ohtani dam (completion in 1993)	Aug. 1969 Flood

an incident that led to the construction of the Kasahori dam in 1964, which permitted the design scale of the river to be revised up to 1,600 m^3/sec. Then after disastrous flooding in August 1969, the Otani Dam was completed in 1993, which allowed revision of the river's design scale to 2,400 m^3/sec, and the basic high-water flow capacity to 3,600 m^3/sec (Niigata Prefecture data, 2005).

Work began on the Kariyata River after a flood in 1949, and even more substantial public work was done on the Kariyata River after a disastrous flood in 1964, at which time the Kariyata dam was planned. The design scale of the river was revised upward several times to 950 m^3/sec, 1,050 m^3/sec, and finally to 1,550 m^3/sec with a basic high-water flow capacity of 1,700 m^3/sec when the dam was constructed. Through these repeated public works improvements, the incidence of flooding on the Ikarashi and Kariyata Rivers was greatly reduced, communities were spared common flooding, and the risk of rains heavy enough to cause flooding was reduced to about once every 100 years. But the possibility of large-scale flooding that exceeded the design scale of the local rivers was never entirely eliminated. Indeed, as we shall describe, a scenario emerged of a new potential risk of substantially larger-scale hazards as a result of levee failure.

3.1.2 Intangible damage mitigation initiatives

It is apparent though interviews with local officials that the main thrust of preliminary disaster preparedness in the stricken area was to rely on large-scale public infrastructural improvements -continuous high levees and dams- for flood hazard control, and not enough concern was given to intangible flood mitigation efforts. Earlier we described how there was little awareness among local officials and the local populace about the risk of flooding, and it is apparent from our interviews and from newspaper accounts, that genuine advance preparedness for flooding by the government and the local people was inadequate. For example, in Nakanoshima Town "it was known that the waters of the Kariyata River had reached dangerously high levels two hours before the levees failed, but they stepped up inspections and, based on past experience, issued an evacuation advisory just before the levees actually failed, saying that the situation was under control". The fact that the town offices were inundated is another indication that advance flood preparation was insufficient.

3.2 Local and citizens' response: Efforts by the local fire brigade and volunteers

Although the local populous was only dimly aware of the flooding risk, the Volunteer Flood Brigade (organized on a community basis) played a major role in mitigating hardship and damage from the flood. Beginning in the early morning of 13 July 2004, there was one catastrophe after another in Sanjo City: slope failures, inland flooding, rising water level in rivers, and the waters of the Ikarashi started coming over the levee. In the midst of efforts to respond to all these crises by the city, the Fire Department, the Volunteer Flood Brigade, and others, at 1:15 in the afternoon thc Ikarashi River levee failed. First water started coming over the top of the levee at 12:40, then 10 minutes later the back of the levee began to be eroded. By 1:00 a 50-m breach had opened up and a muddy torrent of water was pouring through, and at 1:25 the levee failed and a huge breach 117 m across was washed out.

By nightfall, the entire Rannan District was covered with water up to 2 m deep in some places. Rescue efforts began immediately after the levee failed with a total man-days of 4,500 people involved in the effort, including volunteer Flood Brigade personnel from nearby cities and towns, members of the Self Defense Forces, coast guard personnel, policemen, and others. Close to 40% or 1,800 of these rescue workers were members of the Volunteer Flood Brigades, from Sanjo City itself and other nearby cities and towns (see Fig. 3-1).

By the morning of 15 July the waters had receded, and by nightfall the emergency call-out of regular and volunteer flood fighters was cancelled. Then the cleanup efforts started when 29,000 volunteers descended on the disaster-stricken area to help the victims clear mud and debris from their homes.

At the initial stage of seeing multiple crises unfolding across the entire city at the same time, the government realized that it was going to take substantial human resources to cope with the situation. Considering that the Ikarashi levee had already sprung leaks, overtopped, and failed in 13 places, it was

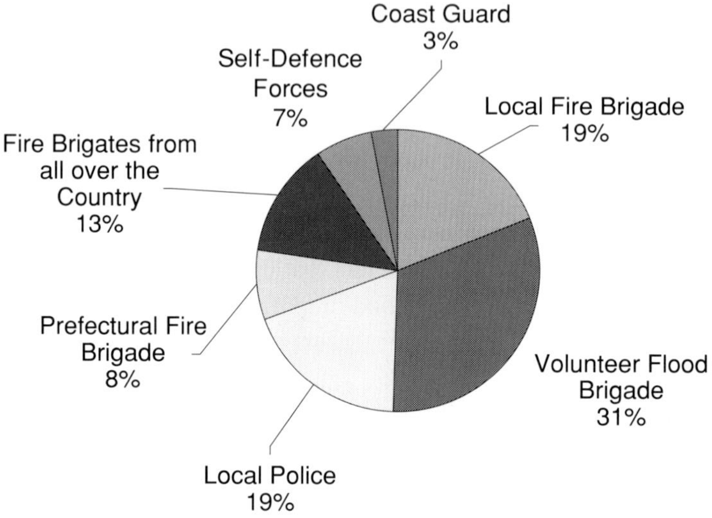

Fig. 3-1. Flood fighting and rescue operations by disaster prevention related agencies in the Sanjyo City.

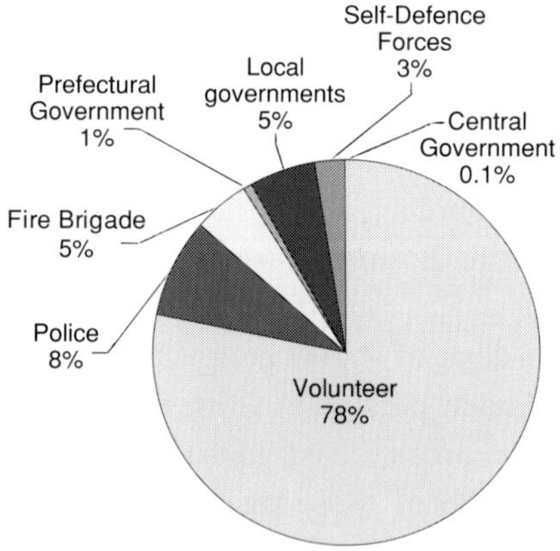

Fig. 3-2. Recovery operations by disaster prevention related agencies in the Sanjyo City.

entirely possible that the levee could fail elsewhere, so in addition to the regular firefighters employed by the local government who numbered only 152, the government also mobilized 1,166 workers from the volunteer Flood Brigades (7.7 times the number of regular firefighters). Massive human resources were committed to the flood control efforts that were needed at many different places at the same time, and the additional manpower provided by the

Photo 1. Flood-fighting activities to prevent water leakage of the levee.

Volunteer Flood Brigades made up of local communities was indispensable. Photograph 1 shows the flood control activities in progress.

After the disaster had occurred, Flood Brigade volunteers from throughout the prefecture and beyond were quickly mobilized through extensive government coordination, and rushed to the scene to help fight the flood and assist the victims. The involvement of the National Disaster Volunteer Center that organized 30,000 volunteers to come in and help with the cleanup and restoration in the aftermath of the flood, also deserves special mention (Fig. 3-2).

4 Extent of Devastation

During the Niigata flood of 2004, levees failed at 11 different places on six different small-to-medium-sized rivers throughout the Chuetsu region in the Niigata prefecture, and more than 340 mudslides were counted throughout the prefecture centering again on the Chuetsu region. In the Ikarashi River drainage area, this resulted in the inundation of 490 ha of residential neighborhoods and 830 ha of farmland, and in the Kariyata River basin, 250 ha of residential land and 903 ha of farmland were flooded, for a total of 2,473 ha of farm and developed land covered by the flood waters. In Sanjo and Mitsuke Cities 2,914 non-residential buildings were damaged or destroyed, a number far in excess of the 1,776 building destroyed in the previously most destructive flood. Sanjo suffered the greatest number of casualties with nine deaths caused by the flood, and 12 of the victims were elderly people in their 70s.

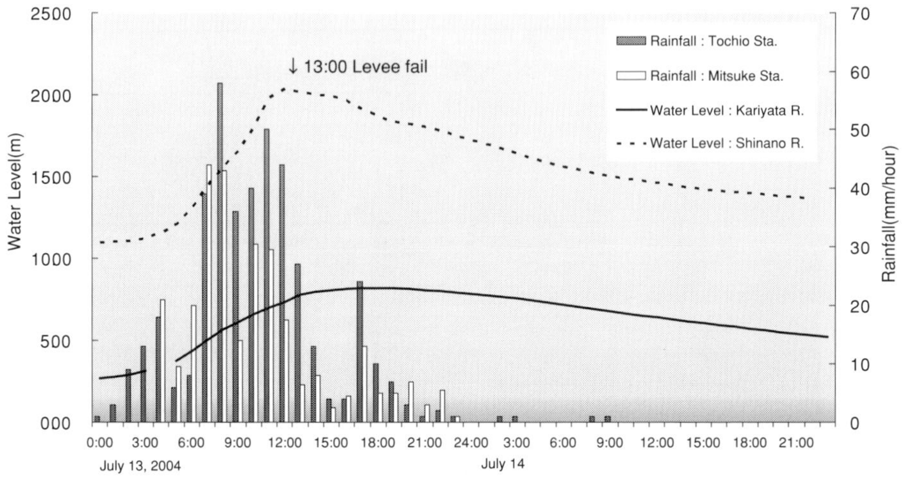

Fig. 4-1. Hydrographs of the Kariyata and Shinano Rivers and hourly amount of rainfall.

The total financial losses resulting from the disaster were ¥200.125 billion (Niigata Prefecture Public Works Department River Management Section, 2005), of which 98% was attributed to flooding (¥196.1225 billion) and 2% attributed to mudslides (¥4.0025 billion). Breaking down the economic losses by sector, 71.62% (¥100.545 billion) of the damage was to the private sector property, 0.44% (¥900 million) to public utilities, and 27.94% (¥57 billion) to public works infrastructure.

Turning now to loss density, we find that dividing Sanjo's private sector economic losses of ¥82.3 billion by the total number of residential hectares in Sanjo, that were inundated (490 ha), we obtain a loss density (i.e., private losses per ha) of ¥170 million per ha. Nationwide, we have been observing a trend toward increasing loss density in recent years. The devastation of commercial parts of Nagoya by the Tokai flood in 2000 when the Shin River levee failed, resulted in a loss density of ¥121.583 million per ha (Sato, 2002). The fact that the loss density of the Sanjo City flood tops even that figure, reveal just how destructive the Sanjo flood was.

Along with the structural damage, much of what people had inside their homes-tatami mats on the floor, furnishings, and so on-was ruined and added to the vast amount of flood-damaged refuse that was discarded. These mountains of water-damaged garbage hampered cleanup efforts in the flooded neighborhoods, and concerns over sanitation and environmental contamination also slowed down the restoration work and had other detrimental effects.

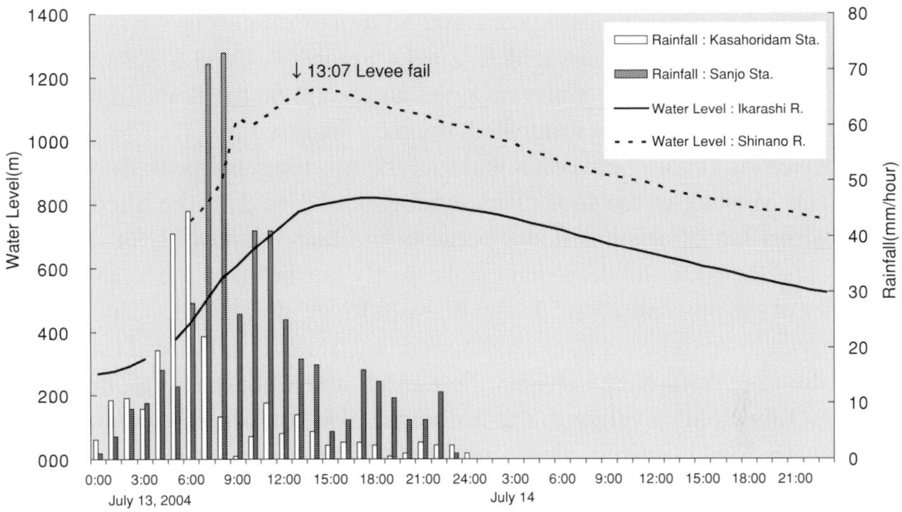

Fig. 4-2. Hydrographs of the Ikarashi and Shinano Rivers and hourly amount of rainfall.

A month after the flood, the local newspaper reported that, on 12 August close to 48,000 tons of garbage resulting from the flood had been collected at three makeshift sites including the former Sanjo horse track, and it was going to cost more than an estimated ¥2.6 billion to dispose of the garbage (*Niigata Nippo*, 17 August, 2004).

5 Characteristics of the 2004 Niigata Flood Hazard

5.1 Rain

The cause of the 2004 Niigata floods was the inordinate amount of rainfall—both the total amount and rainfall over a short period of time—that was intensified by the seasonal rain (baiu) front that stalled over the region. The rain extended over the entire lower reaches of the Shinano River, and was especially intense over the drainages of the Ikarashi and Kariyata Rivers in the Chuetsu region. The headwaters of both rivers received 200–400 mm of precipitation over a 24-hour period (see hourly rainfall for the Kariyata River drainage in Fig. 4.1).

Reconstructing torrential rain patterns over different intervals based on precipitation data collected over a 23-year period (1979–2002) by the AMeDAS (Automated Meteorological Data Acquisition System) rain observation station at Tochio, Niigata, it was found that 422 mm of rainfall over a 24-hour period would occur every 530 years, 267 mm of rainfall over a 6-hour period would occur every 500 years, and 58 mm of rainfall in a 1-hour period would occur every 208 years (JSCE, 2004). Figures 4-1 and 4-2 show hydrographs for Mitsuke on the Kariyata River and Sanjo on the Ikarashi River, respectively. In the Kariyata watershed, the most intense rainfall of 58 mm/hour fell upriver on Tochio between 8 and 9 in the morning, but peak flooding did not reach Mitsuke at the foot of mountains until 1 to 2 in the afternoon, so there was a lag of about 4 hours between the heavy rainfall over the headwaters and the peak flooding that can be partly attributed to the water storage capacity of the upstream dam. In the Ikarashi River, the peak flooding reached the alluvial lower reaches after an 8-hour delay. If the flood arrival time is two times the time difference between the rainfall and when the peak flood flow appears, this yields a value of 8 to 16 hours. Thus the amounts of rain when the Niigata flood occurred were equivalent to a precipitation event that might occur once in approximately 500 years, and this concentrated torrential rain caused the water levels recorded at many of the observation stations along the rivers to break all previous records.

5.2 Conditions exacerbating the flood hazard

Typically, flood waters in the lowlands downriver from the Ikarashi and Kariyata Rivers flow slowly because the slope is fairly gradual, but the scale and character of the 2004 Niigata flood were vastly different because the removal of structural impediments when the levees failed resulted in a powerful flood flow that swept through the breach to cover an immense area.

5.3 Increased volume of inland flooding due to levee failure

Approximately 13.93 million m^3 of water poured through the breach in the left bank levee of the Ikarashi as opposed to only 340 thousand m^3 of water that came over the top of the levee in that sector. It is thus clear that the overwhelming bulk of flood waters that inundated the inland area (about 98% of the total) can be attributed to the failure of the levee. In other words, the failure of the levee resulted in a sudden surge of massive amounts of flood water, so that a much more extensive area was inundated with deeper water, and the force of the flood water surge was extremely powerful around the breach.

Photograph 2 shows the breach in the Kariyata River levee in Nakanoshima at 6:11 pm. The Buddhist temple Myoeiji that stood close to the breach has already been swept away, and houses in the area are submerged up to second floor level. Levees are intended protect the inland areas and communities behind them, but when they fail they expose these same communities to massive devastation from surging flood waters.

Photo 2. Flood water flowing into the housing area from the bank break site of Kariyata River (Photo by Kyodo-Tushinsha).

5.3.1 Increased flood flow hydrodynamic force due to levee failure

The flood plain around where the levee failed slopes very gradually at a rate of about 0.7 to 0.8 meters per 1,000 m, so the flood waters that came over the top of the levee flowed very placidly. But the flood waters that surged through the breach in the levee were extremely powerful, indeed so powerful that many homes located even considerable distance from the breach were completely destroyed within a very short time. Members of the 7.13 (13 July) Niigata Rain and Flood Disaster Survey Committee charged with investigating the flood have discussed at length whether this incredible surge force of the water through the breach can be attributed to the difference in water level between the river channel and the lowlands behind the levee, or to the abruptness with which the levee failed (Niigata Prefecture).

5.3.2 Increased flood peak flow and infrastructure improvements

Earlier we have described how flooding on the Ikarashi and Kariyata Rivers and the history of efforts to cope with the flooding exhibit a classic pattern of flooding⟶flood control improvements⟶more flooding⟶more flood control improvements⟶levees fail⟶massive flooding. Certainly, these repeated public works infrastructure improvements have enhanced flood control safety, reduced the incidence of flooding on the Ikarashi and Kariyata Rivers, and spared downriver communities from common flooding. Yet we must also realize that straightening out meandering river channels and building long continuous levees to inhibit flooding upriver can have the adverse effect of enabling substantially greater flood flows on the lower reaches of the river. We would also note that infrastructure improvements increase the volume of flood flow waters that course through river channels by 2.1 to 2.4 times, and considering that the difference in elevation between the water level of flooded rivers and the inland areas behind the levees in downriver areas is close to 5 m, people must be prepared for much larger-scale flooding in the event that levees do fail.

6 Summary

A major flood disaster occurred in Niigata Prefecture on 13 July 2004 as a result of heavy rainfall that exceeded the design scale of the river infrastructure. Flood waters breached the levees in downstream urban areas, causing tremendous damage. Examining the flood from the perspectives of risk theory, we observe the following specific characteristics:

1. The flood was of a low probability but high consequences (LPHC) type; that is, an event that rarely occurs, but results in catastrophic damages when it does occur.
2. Factors contributing to increased flood hazard were (1) levee failure due to flood waters far exceeding the design scales of the rivers, which increased the force and volume of the flood waters, and (2) development of a narrow valley plain without considering the potential for LPHC-type flooding.
3. Private sector economic losses were substantial at ¥154 billion, representing 71.62% of the total economic losses. This is the second highest percentage of private sector losses in Japan's history, only exceeded by a loss figure of 80% in the Tokai flood that devastated urban areas of Nagoya in 2000.
4. In terms of human suffering, a large number of the casualties were elderly. This could be seen as a reflection of the aging of Japanese society.
5. Regional disaster prevention plans and flood prevention plans created by local governments consider the possibility that levees might fail in the event of a major flood. However, local officials in charge of flood-disaster management had no true sense that such an event might ac-

tually occur, so advance measures and preparations were insufficient. Local residents also showed little awareness of the danger of LPHC type flooding, and almost none had made any advanced preparations.

6. Locally organized Flood Brigades played a major role in efforts involving large numbers of people mobilized to undertake rescue and recovery activities. Indeed, these local Flood Brigades were on the scene almost immediately after the flood occurred, and until support started to arrive from other regions, they accounted for 30% of the total number of people on the ground working on behalf of the regional disaster prevention organizations.

We should also note that, according to some reports, Flood Brigade efforts to stem overtopping by piling sandbags on the levees was a key factor in the eventual failure of the levees, which of course was the primary cause of the devastating inland flooding.

Materials and Data

I Local Government Materials

1) Niigata Prefecture, materials relating to the July 2004 Niigata flood disaster.
2) Mitsuke City, materials relating to the July 2004 Niigata flood disaster.
3) Nakanoshima Town, materials relating to the July 2004 Niigata flood disaster.
4) Sanjo City, materials relating to the July 2004 Niigata flood disaster.

II Government Institute Materials

1) Ministry of Land Infrastructure and Transport River Bureau, materials relating to the July 2004 Niigata flood disaster, http://www.mlit.go.jp/.
2) Ministry of Land Infrastructure and Transport Hokuriku Regional Development Bureau, materials relating to the July 2004 Niigata flood disaster, http://www.hrr.nilit.go.jp/.
3) Cabinet Office, Government of Japan, materials relating to the July 2004 Niigata flood disaster.
4) Fire and Disaster Management Agency, materials relating to the July 2004 Niigata flood disaster.
5) Niigata District Weather Station, weather information relating to the July 2004 Niigata and Fukushima flooding (final report), p. 25.

III Newspaper Articles

1) Articles in *Niigata Nippo News*, July 14 to September 30, 2004.

IV Maps

1) Geographical Survey Institute maps.

1:25,000 scale, Mitsuke City, source revised and published in 1948, updated and revised in 2001.

Sanjo City, revised in 1931, source revised and published in 1948, updated and revised in 2001.

1:200,000 scale, Niigata and Nagaoka 1996.

1:25,000 scale, topographical map, p. 11.

2) Geographical Survey Institute, 1:25,000 scale, Sanjo, topographical map, p. 11, 1990.

3) Nakanoshima district planning map.

4) Mitsuke City planning map.

5) Sanjo City planning map.

References

Japan Society of Civil Engineers (JSCE) (2004): "July 2004 Hokuriku Heavy Rain Disaster Emergency Investigation Commission materials", August 31 (in Japanese).

Hokuriku Construction Association (2004): Hotto Hokuriku, July 2004 Special Issue, No. 43 (in Japanese).

T. Sato (2005): "Flood Risk Structure and Characteristics: Toward an Integrated Flood Risk Management Framework", *Keio University Hiyoshi Review of Social Sciences*, 15, pp. 25–38 (in Japanese).

T. Sato and K. Nakane (2004): "Course of Events of Sanjo City Flood Disaster", *Chiri*, Vol. 49, No. 12, Kokon Shoin, pp. 28–31 (in Japanese).

T. Sato, K. Seo, S. Ikeda, and T. Nagasaka (2005): "Cooperation Between Local Governments and Disaster Prevention Organizations in the 2004 Chuetsu Earthquake", Chuetsu Earthquake Disaster Survey Report, a urgent research project supported by special coordination funds for promoting science and technology (in Japanese).

T. Sato (2002): "Time and Spacial Distrubution of the Damage in the 2000 Tokai Urban Flood Diasters", Natural Disaster Research Report, No. 38, National Research Institute for Earth Science and Disaster Prevention (NIED), pp. 99–161 (in Japanese).

T. Sato, K. Nakane, S. Ikeda, and T. Nagasaka (2006): "Characteristics of the Flood Risk of the 2004-Niigata Flood Disasters", Natural Disaster Research Report, No. 40, National Research Institute for Earth Science and Disaster Prevention (NIED), pp. 33–78 (in Japanese).

A better integrated management of disaster risks: Toward resilient society to emerging disaster risks in mega-cities, Eds., S. Ikeda, T. Fukuzono, and T. Sato, pp. 193–198.

Insurance Issues of Catastrophic Disasters in Japan: Lessons from the 2005 Hurricane Katrina Disaster

Hiroaki Tsubokawa

1 Introduction

A series of hurricanes including Hurricane Katrina hit the Gulf Coast in the summer of 2005, causing the worst damage in the U.S. history of natural disaster and raising many issues regarding the nation's insurance system. The experience of Hurricane Katrina can provide valuable lessons for Japan. It must be remembered that this kind of disaster could also happen in the future in Japan, which has a great deal of risk for major disasters such as large-scale typhoons, localized torrential rains, and massive earthquakes. In this paper, I will present an overview of the events of Hurricane Katrina and discuss the lessons to be learned in Japan.

2 Natural Disasters and Problems of Public Insurance Systems

Founded in 1968, the National Flood Insurance Program (NFIP: http://www.fema.gov/nfip/), administered by the Federal Emergency Management Agency (FEMA), is a state-run insurance system with a history of nearly 40 years. In general, homeowners insurance from the private sector does not cover flood damage. A geographical bias is seen with NFIP policies, as is usual with insurance for natural disasters. A hurricane-prone region extends from the southern United States to its eastern seaboard, and about 40% of NFIP policies are concentrated in Florida, followed by Texas in second place. Louisiana has the third highest number of policies, accounting for about 8% of all NFIP policies. Many policies in Louisiana are located in highly flood-prone areas such as New Orleans and Jefferson Parish. One reason for the difficulty of administering natural disaster insurance is that policies are concentrated in high-risk areas, resulting in higher premiums. In Japan as well, there are clear differences in the number of earthquake insurance policies in each region. Most policies are concentrated in the Tokyo metropolitan area and the prefectures of the Pacific seaboard (Tsubokawa, 2004). Table 1 presents a comparison of NFIP and Japan's earthquake insurance (http://www.nihonjishin.co.jp/english/index.html), the major insurance programs with public support in the U.S. and Japan. In Japan, unlike the U.S.,

Table 1. Comparison of NFIP and Japan's earthquake insurance.

Name	National Flood Insurance Program (NFIP)*	Earthquake Insurance**
Year created	1968	1966
Administered by	Federal government	Private sector and national government
Role of private insurance firms	Sales. Through the "Write Your Own" (WYO) program, private insurance companies can handle flood insurance sales in the same way as their own products. Most policies are sold through WYO.	Sales and primary insurance. All primary insurance policies are ceded to the Japan Earthquake Reinsurance Co., Ltd. for reinsurance, and the national government provides reinsurance.
Role of national government	Insurer	Reinsurer (excess of loss reinsurance)
Number of insurance policies	4.56 million policies (as of December 2004)	9.32 million policies (as of March 2005)
Total coverage	$743 billion	¥71 trillion
Total annual premiums	$ 2 billion	¥120 billion
Premium rate categories	Divided into about ten zones, based on flood hazard maps prepared in accordance with FEMA surveys.	Four zones nationwide, plus two classifications based on building structure (wood or other).
Premium discount system	Under the Community Rating System (CRS), discounts of up to 45% are available based on community efforts to reduce risk.	Discounts of up to 30% are available, depending on the building's earthquake resistance and year of construction.
Highest claims settlement incident	Hurricane Katrina, at least $20 billion (estimated)	Hanshin-Awaji (Kobe) Earthquake, ¥78 billion
Maximum total payment	No particular limit is set. However, following Katrina, the limit on borrowing from the federal government was raised from $1.5 billion to $3.5 billion.	¥5 trillion. Under law, if the total exceeds ¥5 trillion, benefits will be proportionately reduced.
Participation rate	Varies by region, ranging from over 70% in some areas to a few percentage points in others.	Varies somewhat by region. Nationwide, 37.4% of fire insurance policy holders also had earthquake insurance in 2004.
Recent developments	The Flood Insurance Reform Act of 2004 revises compensation for properties subject to repeated claims. Disaster victims receiving benefits in multiple incidents are subject to revised compensation.	Study is underway concerning introduction of a system to discount insurance premiums for existing buildings as well, reflecting the results of earthquake resistance evaluation by local governments and the like. Income tax deductions for earthquake insurance premiums are also being considered as a way to promote adoption of earthquake insurance.

*NFIP: http://www.fema.gov/nfip/
**Earthquake Insurance: http://www.nihonjishin.co.jp/english/index.html

flood damage is covered by private insurance alone, due to the country's past experiences and flood control characteristics. However, earthquake risk involves the potential for large-scale natural disasters throughout Japan, so it is necessary for the national government to provide support in the form of reinsurance. Some problems with NFIP have been pointed out following the hurricane disaster.

First, it is difficult to rebuild a house with NFIP insurance benefits alone.

Insurance coverage is limited to the primary dwelling, and there are limitations on insurance benefits for the building and household goods. The approach of placing limits on insurance benefits is often used in public insurance systems. For example, there are limited benefits in Japan's earthquake insurance, which only pays up to half of the benefits from fire insurance (NLIRO, 2003). This limited coverage is one reason why homeowners may hesitate to purchase insurance from NFIP. Homeowners may also purchase insufficient coverage because of high insurance premiums, resulting in inadequate benefits in case of a claim. According to FEMA, the average insurance premium with NFIP is $438 per year. For a nonprofit public insurance program that only covers flood damage, this is quite expensive. Homeowners hesitate to purchase flood insurance because of the high premiums and limited coverage.

Second, NFIP does not cover miscellaneous costs associated with a flood, such as temporary living expenses. "Not having that coverage is going to bankrupt a lot of people," said Alex Soto, an insurance agent in Miami. According to Edward Pasterick, spokesman for NFIP, "Covering additional living expenses would be very expensive. It might put the price of the coverage out of reach." (Both quotes from www.sun-sentinel.com, "Floridians keep eye on program as reform urged," October 9, 2005.) Japan's earthquake insurance also consists only of damage insurance and does not cover any incidental expenses. Proposals have been made several times in the past to introduce coverage for incidental expenses, but as in the case of NFIP, this was abandoned because increases in insurance premiums would be inevitable.

Third, there are issues related to the need for flexibility and more efficient handling of claims settlement. One troublesome point in the administration of a natural disaster insurance program is that a large number of claims occur simultaneously. It is important to handle claims settlement promptly, but there are limits to the extent of advance preparation that can be done, and the costs are considerable. Although a large number of claims adjustment officers were dispatched from all over the U.S., New Orleans remained underwater for a lengthy period of time, and it was initially a slow process to check on damaged properties. As disaster victims demanded a faster and more efficient claims settlement response, FEMA decided to use aerial photographs to reach a decision about properties located within certain areas, without requiring on-site confirmation. In the kind of massive ocean-trench earthquake that is expected to occur in Japan in the future along the Tokai-Nankai Trough, an unprecedented number of claims would be inevitable. The Katrina experience can serve as a valuable reference.

Last, issues related to risk communication can, in a sense, be seen as the most important problem area. On October 17, the *Washington Post* reported that in the Lower Ninth Ward of New Orleans, which suffered heavy flood damage, few residents had purchased flood insurance because this area had been evaluated as low-risk in a flood insurance map prepared by FEMA. (*Washington Post*, "Risk Estimate Led to Few Flood Policies: For Most in New Orleans' Ninth Ward, Extra Coverage Wasn't Required," October 17, 2005.) Mortgage companies generally require NFIP policies when taking out a home loan, but there was no such requirement in this area.

This reality is not only a problem of insurance, but also involves important problems regarding risk evaluation for low-frequency events as well as the expression and interpretation of such evaluations. In this case, the results of risk evaluation were used and interpreted in a manner that ended up having the opposite effect to its intended outcome of reducing risk.

3 Debate on the Accuracy of Claims Settlement Models

A Swiss reinsurance company has estimated the total damage from natural disasters worldwide in 2005 at $225 billion. Damage from Hurricane Katrina is estimated at $135 billion, or 60% of the total (Swiss Re.: http://www.swissre.com/).

Forecasting techniques have reportedly become more accurate and sophisticated due to advances in computer technology in recent years. Damage estimates are calculated with predictive functions based on variables regarding the strength of a disaster event and the vulnerability of property subject to damage. In general, wind speed is the central factor in the destructive force of a hurricane or typhoon, and wind speed is generally used as the index expressing damage strength. However, this technique alone cannot produce accurate estimates when levees are breached and flooding also enters the picture, as in the Katrina disaster.

Market trust for modeling technology has reportedly declined because of the wide range of damage forecasts, from several billion to several hundred billion dollars. This subject was discussed at a reinsurance conference held in Monte Carlo in September 2004 (*Business Insurance*, "Modeling Helpful but no Substitute for Underwriting," September 26, 2005). As simulations become more sophisticated, the reproducibility of physical phenomena has greatly improved. Meanwhile, information concerning the damage involves elements that are difficult to quantify regarding the attributes of buildings, household goods, and other insured property, and there are limits to the level of accuracy that can be achieved. The insurance industry is gaining a better understanding of the limits and uses of modeling, but it is dangerous for underwriting to rely on modeling alone.

Another basic problem in the future modeling of natural disasters is that opinions are divided on predictive elements for climate change and large-scale abnormal events. More highly accurate research is needed regarding the effects that global warming due to increases in carbon dioxide levels will exert on catastrophic events. For the determination of probable maximum loss (PML), an important index in insurance operations, further debate is also needed regarding future predictions and the probability of detection of large-scale natural disasters that have not yet been experienced.

4 The Role of Insurance in Societies where Inequalities Exist

NFIP is a special type of insurance with a highly public nature, administered by the federal government. Nonetheless, as stated earlier, its insurance premiums are certainly not cheap. With natural disasters such as floods, another dilemma is that low-income residents who are unable to afford high insurance premiums live in high-risk areas, and as a result, the system cannot function effectively when a disaster occurs. No matter how excellent an insurance product may be, it is meaningless unless it is available. Even if a policy is obtained, no one will purchase continuing coverage unless it is affordable. The keys to the widespread adoption of insurance are availability and affordability. In New Orleans, a vicious cycle has emerged in which poor people cannot obtain insurance benefits and are unable to return to the neighborhoods that were their homes, resulting in further delays to reconstruction.

Insurance is a mirror that reflects a country's society. An excellent insurance system should disperse risk in an effective and impartial manner. The time has come when we must consider how to keep the future of Japan from becoming like the difficult reality faced by American society, with its large inequalities.

5 Summary

Every year, losses from natural disasters continue to increase in the worldwide insurance market. Along with the risks of terrorism and widespread infectious diseases such as avian influenza, natural disasters remain a serious threat for the non-life insurance business. With disasters of this scale, a surprising number of points in common can be found in a variety of subsequent discussions, going beyond the type of disaster or differences among countries. After the Hanshin-Awaji (Kobe) Earthquake, the immunity of fire insurance from earthquake claims was contested in court, and proposals from many quarters called for the creation of a funding system for natural disasters.

Media coverage following Hurricane Katrina was reminiscent of the situation in Japan just 11 years ago. We have many lessons to learn from Hurricane Katrina.

The Kobe Earthquake gave impetus to the preparation of various natural disaster hazard maps in Japan. National Research Institute for Earth Science and Disaster Prevention (NIED) provides probabilistic seismic motion forecasting maps on its website (Japan Seismic Hazard Information Station, http://www.j-shis.bosai.go.jp/), and now everyone can easily find the earthquake risk for their own area of residence. Local governments are working hard to prepare river hazard maps. We have reached an era when each individual will judge his or her own level of risk and make decisions on that basis. The Participatory Flood Risk Communication Support System (PAFRICS) is expected to play a large role in this kind of progress. It is necessary to actively promote risk communication based on the effective use of IT, while taking care to avoid the future development of a digital divide.

References

Non Life Insurance Rating Organization of Japan (NLIRO) (2003): Earthquake Insurance in Japan, 118 pp.

Tsubokawa, H. (2004): Japan's Earthquake Insurance System, *Journal of Japan Association for Earthquake Engineering*, **4**(3), 154–160.

Part III: Pilot studies of implementing social platform of risk management in local community: Participatory flood risk communication support system (Pafrics)

A better integrated management of disaster risks: Toward resilient society to emerging disaster risks in mega-cities, Eds., S. Ikeda, T. Fukuzono, and T. Sato, pp. 199–211.

Participatory Flood Risk Communication Support System (Pafrics)

Teruki Fukuzono, Teruko Sato, Yukiko Takeuchi, Kenji Takao, Shinya Shimokawa, Isamu Suzuki, Guofang Zhai, Kiyomine Terumoto, Toshinari Nagasaga, Kami Seo, and Saburo Ikeda

1 Introduction

A community-based approach is the key to success in integrated disaster risk management. The participation of local people in the design and planning processes in flood risk management, particularly with respect to soft policy measures, is important. For that reason, enhancement of individuals' disaster preparedness and flood risk communication among residents, regional communities, and administrative authorities are needed. In order to promote these activities, we have developed a new participatory type of decision support system that reflects the results of research into the flood risk perception and disaster preventive activities of local people obtained through recent questionnaire surveys. This is called the Participatory Flood Risk Communication Support System (Pafrics). The system facilitates the integration of soft and hard measures in local communities to be resilient against an unexpected scale of flooding. The system has been already used to conduct a kind of social experiment in schools and workshops of disaster prevention by collaborating with volunteer groups of NGO, administrative authorities, and so on. Some examples of use and results are introduced in detail in the following chapter. This provides both configuration and content of the system.

2 Basic Concept for System Development

Pafrics, a flood risk communication support tool, basically aims at popularizing a new concept of flood risk management based on theory of risk analysis. A concept of integrated flood risk management through the combination of physical and procedural disaster control measures taken by administrative authorities, regional communities, and residents will be widely applicable. The system is being developed from the following viewpoints.

(1) Providing support in participatory decision-making.

(2) Defining the responsibilities of administration, local communities, and residents.

(3) Enhancing pubic and local disaster prevention capacity.

(4) Integrating physical and institutional disaster management measures.

(5) Increasing social and economic efficiency, preserving the environment and ecosystems and sharing costs fairly from a long-term viewpoint.

(6) Applying fair and transparent processes.

(7) Providing easy-to-understand flood risk information (estimating qualitatively and quantitatively the possibility of damage and the costs and benefits of damage mitigation).

(8) Developing human resources for promoting new integrated flood risk management.

(9) Building flood risk communication support methods tailored to Japan.

(10) Taking comprehensive disaster prevention measures to build safe and secure local communities.

3 Content of Pafrics

Pafrics is one of the indispensable vehicles for facilitating flood risk governance based on an integrated framework. It helps users obtain a deeper understanding of a new approach to flood risk management and flood control strategies by learning about flood risk or having a simulated experience of the combination of specific measures to reduce the flood risks. The system is designed mainly for small-member workshops administered by facilitators, but can also be used for self-study. The system is therefore designed for use with mobile personal computers. Some of the system functions are already publicly available and the system is accessible via the Internet Web.

3.1 System environment

Requirements for the Pafrics system include the 1) sharing and rapid provision of research results, 2) accessibility via the Internet, 3) easy information update, 4) an edit function for changing the sequence of information displayed and 5) flexible system expansibility. To meet the requirements, the system is composed of contents (information), contents inventory (information catalogue database), and a contents authoring system (editing the sequence of information display). The system environment consists of a database server and web server.

The system profile is given below.

Performance

- Simple mobile system (based on XML. A web server is available).
- Sharing of know-how and intellectual properties of researchers and experts.
- An edit function that provides for easy modification of the sequence of displaying contents and information.
- Easy update of contents.
- System expansibility for reinforcement.

Components

- Contents (information, and still and animated images).
- Contents inventory (information catalogue database).
- Contents authoring (editing the sequence of information display).

Environment

- Database server, web server, and client systems.

3.2 Configuration of Pafrics

The system configuration is shown in Fig. 1. Major functions are (1) flood risk literacy learning support, (2) information provision for participatory flood risk communication support in damage-reduction measures, (3) facilitators support to organize workshops or meetings, and (4) recording of workshop results.

(1) flood risk literacy learning support

The flood risk literacy learning support function (Fig. 2) aims at providing the assistance to users for learning a new approach or concept of flood risk management to make people more knowledgeable about flood risks. The function provides easy-to-understand information on the characteristics of flood risk, a new flood risk management concept, and flood risk assessment methods. It also supports flood risk literacy learning. Pafrics is superior to conventional systems in two respects. First, the system is designed based on a number of our social studies on disaster risks which include public risk perception, relationships among psychological variables such as risk awareness,

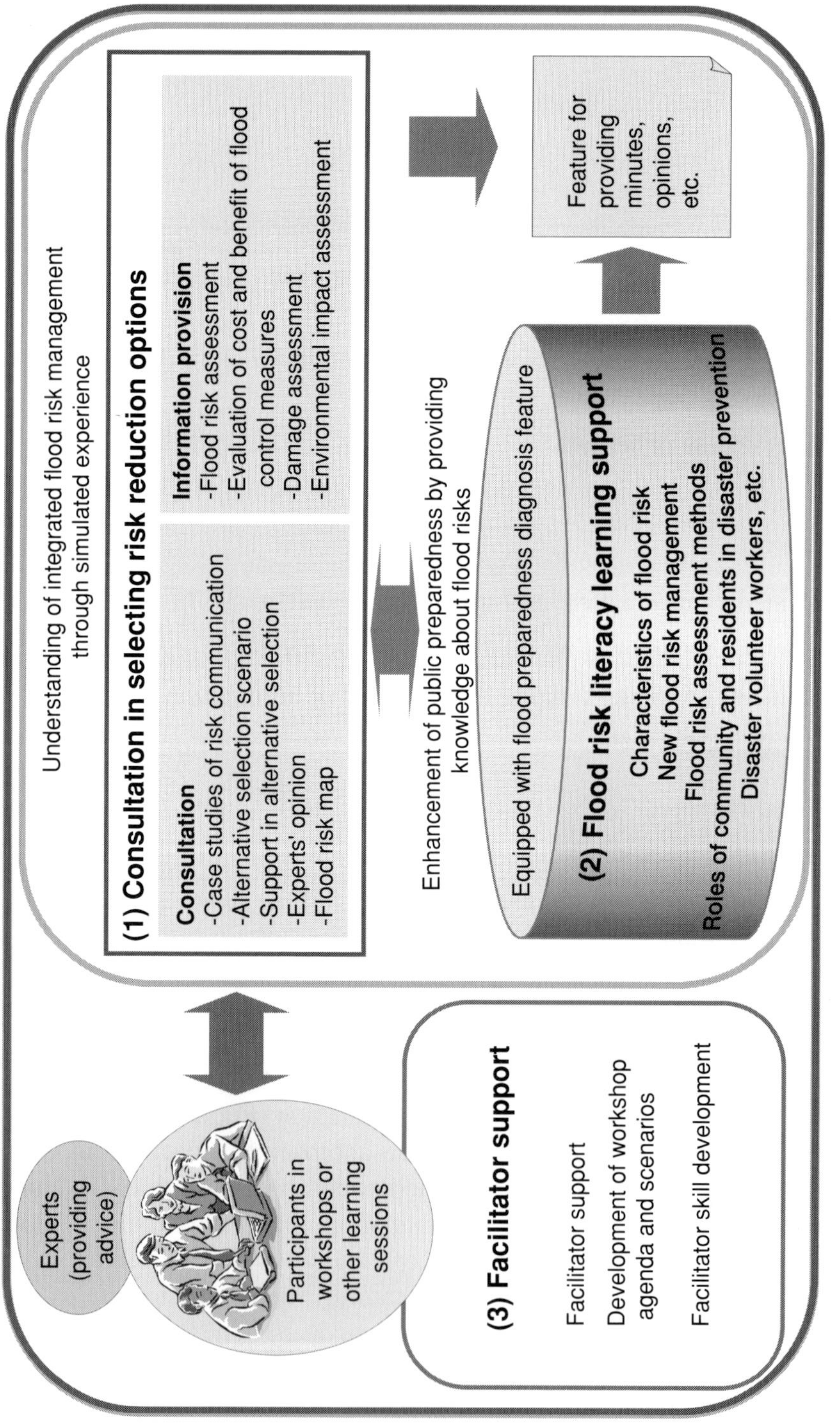

Fig. 1. System configuration.

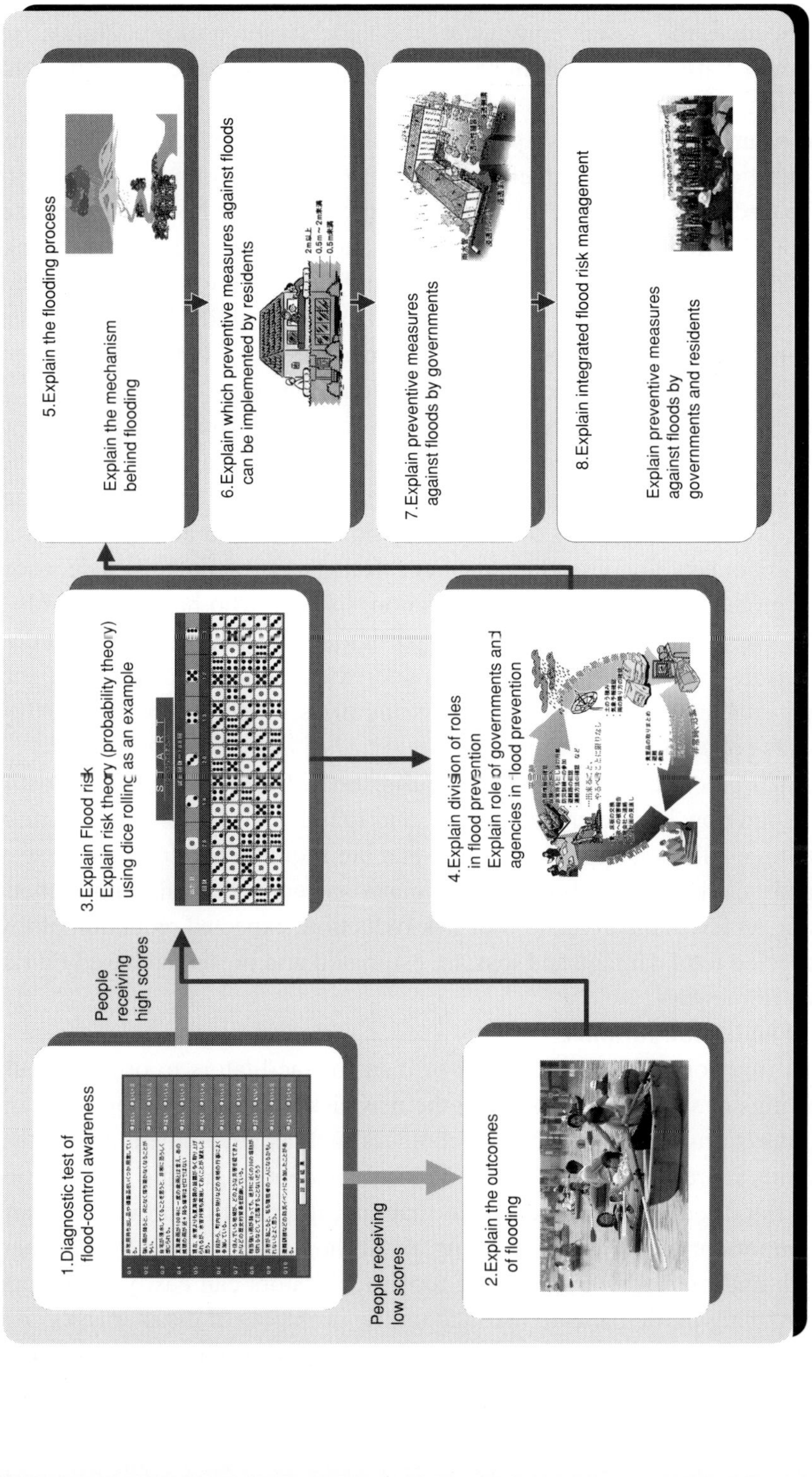

Fig. 2. flood risk literacy learning support function.

anxiety, sense of responsibility, and disaster risk reduction actions (Chapters 7–11). Second, the system emphasizes the understanding of probability, frequency, and statistical distributions of risks, which provides a foundation for considering acceptability of risks (Chapters 4 and 6).

For example, disaster risk preparedness of participants can be measured using questionnaires and contents can be provided according to the results. This is based on a research result that better recognition of flood risk increases self-responsibility, interest in flood damage, and finally the willingness to take flood risk control measures. System users can freely refer to different contents of the database for learning, and use contents according to the "scenario prepared for learning a new flood risk management concept" (Chapters 1–5).

(2) Consultation in selecting risk reduction options

This function provides not only straightforward information relevant to the selection of regional flood control measures, but also objective and detailed information related to flood risk analysis based on scientific knowledge and cost-benefit assessments of risk reduction measures. It also enables users to learn about new flood risk management strategies by simulating a decision-making process based on discussions among people who has diverse values (Fig. 3). The advice sub-function helps participants to select flood risk reduction measures in view of regional characteristics, and the consultation sub-function enables participants to address inquiries to experts regarding their comments concerning assessment results.

Annual precipitation exceeding the threshold, flood flow inside embankments, flood flow behind embankments, and spatial distributions of maximum inundation depths and damage in value that are required for flood risk assessment and cost-benefit analysis, are estimated by conventional, simple methods using an Excel macro code. For risk reduction, expected reduction of loss and expected total damage and loss are estimated and presented based on risk analysis framework.

(3) Facilitator support

This function provides materials or data for facilitators to organize public meetings or workshops regarding the new risk management approach and to lead the discussions between stakeholders such as government authorities, community representatives, NGOs, and local residents regarding the selection of local flood reduction measures. To that end, the function provides a number of important scenarios for learning about the new flood risk management approach and methods. Using these tools, facilitators can easily lead workshops enabling participants to learn about the problems of disaster risks. Another supporting function is also available to facilitators who are familiar with

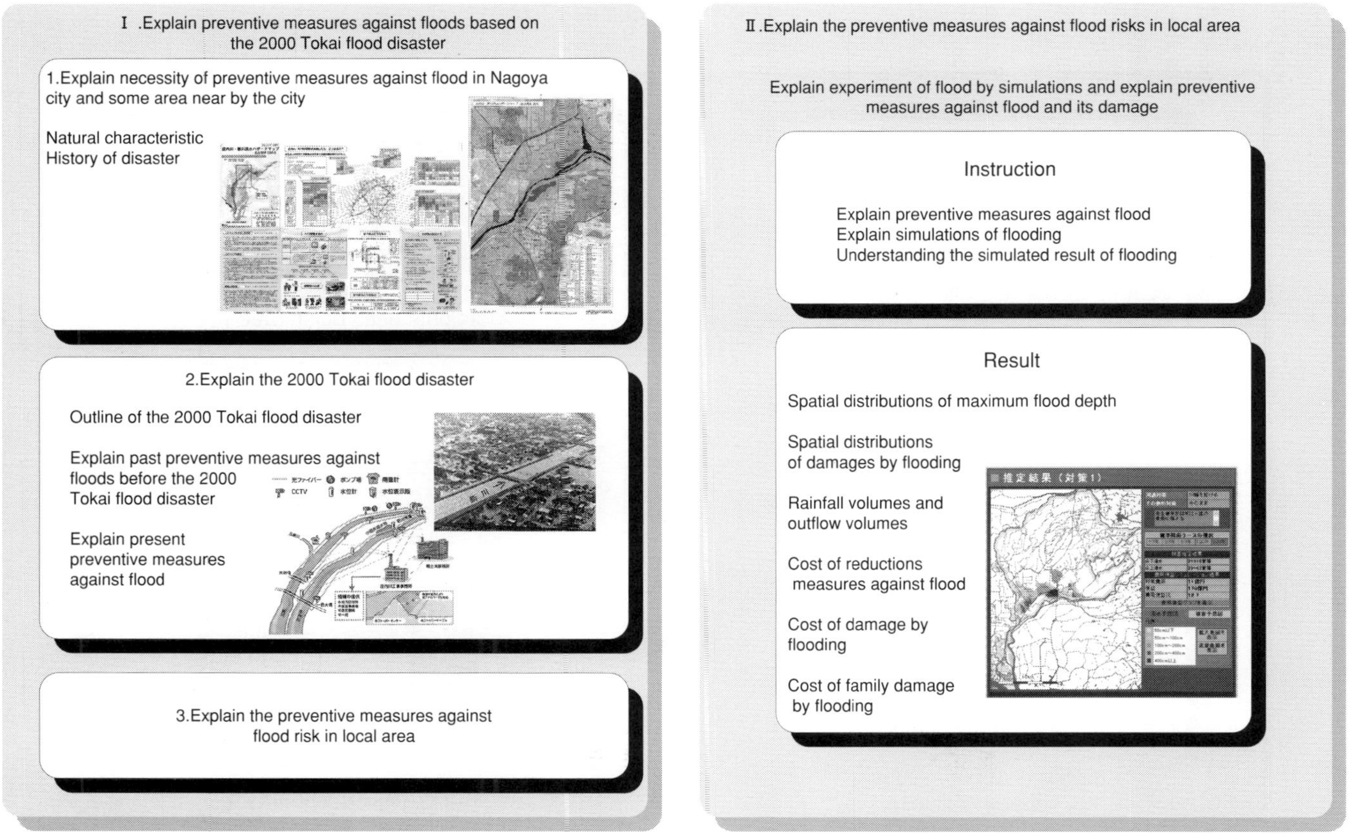

Fig. 3. Support function for consultation in selecting risk reduction options.

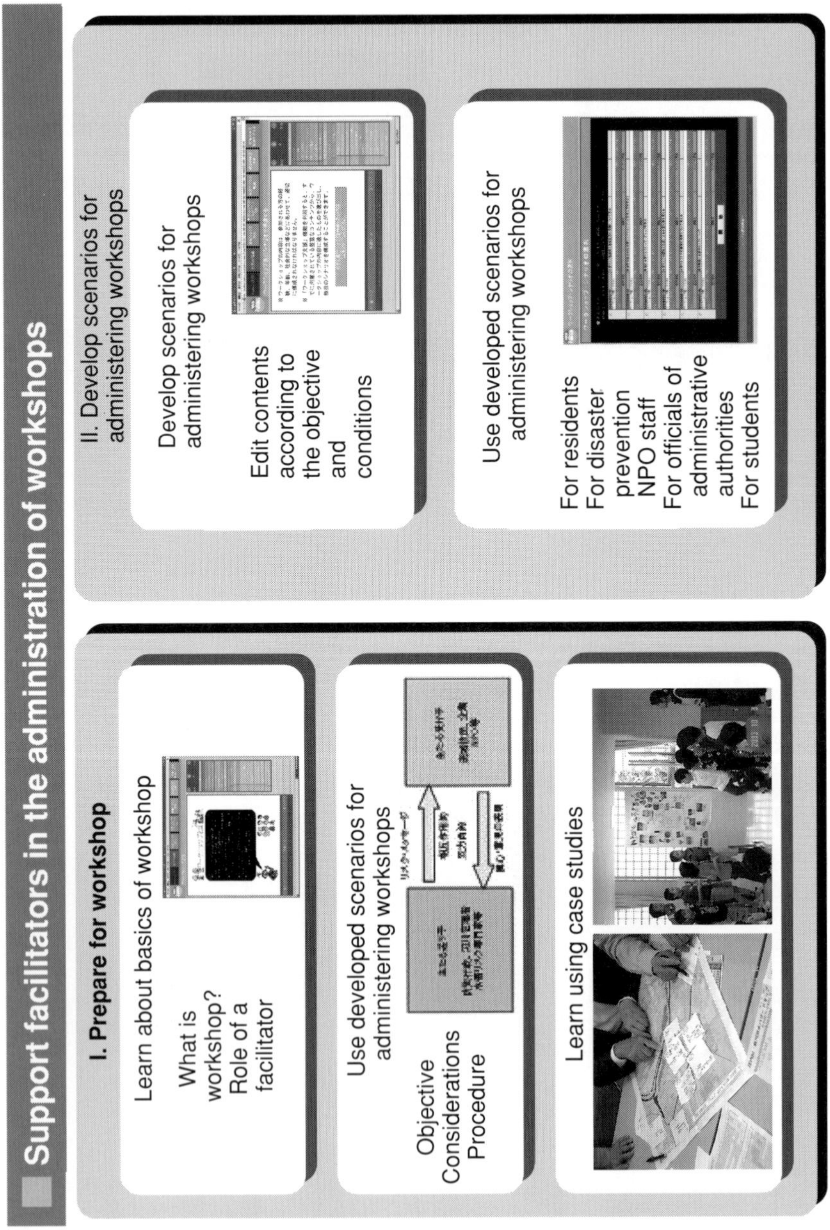

Fig. 4. Facilitator support feature.

Fig. 5. Enter page of Pafrics (Original is in Japanese).

Pafrics. The facilitator can make his own workshop scenario by editing the sequence and content of the information accommodated in the Pafrics or by developing original contents of the scenarios according to the size, duration, and participants of a workshop. The facilitator support function also includes manuals on preparing and organizing workshops, and reporting outcome of the meetings.

4 Pafrics on the Internet Web

4.1 Outline

An Internet-accessible variation of Pafrics has been developed and is now publicly available. Users can access the contents prepared for workshops by simply selecting one of the desired scenarios using a general browser (e.g., Microsoft Internet Explorer). The website consists of two features of Pafrics, "flood risk literacy learning support" and "facilitator support". The "flood risk literacy learning support" feature provides scenarios for operating workshops for teaching participants on flood risks by selecting contents. Descriptions

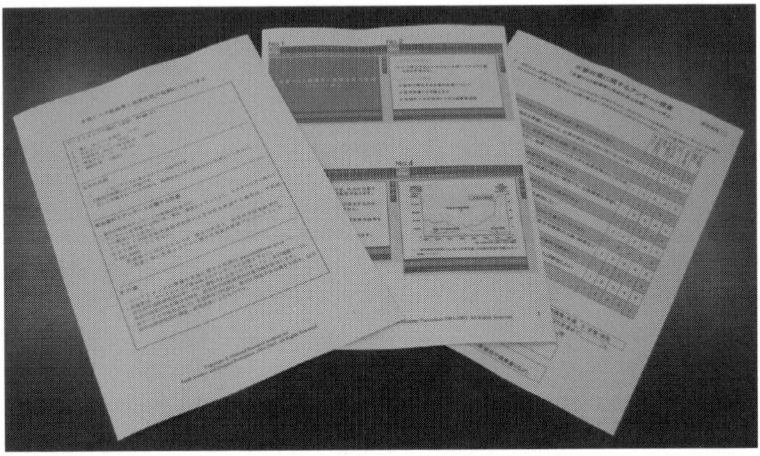

Fig. 6. Documents available at Pafrics website.

of the content for each scenario, guidelines on workshop operation methods, and questionnaires to be distributed to workshop participants are provided to organizers and facilitators of workshops. In the future, it will be made possible for users to register their own content to create new workshop scenarios. So far, we have installed the following six scenarios on the web site (http://www.bousai.go.jp/sougou/shakai/index.html) to have wider public accessibility to the social platform of disaster risk communication support system:

(1) Flood risk reduction measures and the role of local residents.

(2) Concept of integrated participatory flood risk management.

(3) Local flood risks and hazard maps.

(4) flood risk awareness and flood control measures taken by residents.

(5) Characteristics of flood risks and damage.

(6) Probable precipitation and flood risk.

Each scenario is designed to take 60 minutes to be given at workshops. Descriptions, material to be distributed, and questionnaires to be distributed to participants are provided in a package. Terminologies, references and questionnaires to be distributed to organizers are also available for facilitator learning support and information gathering for enhancing Pafrics.

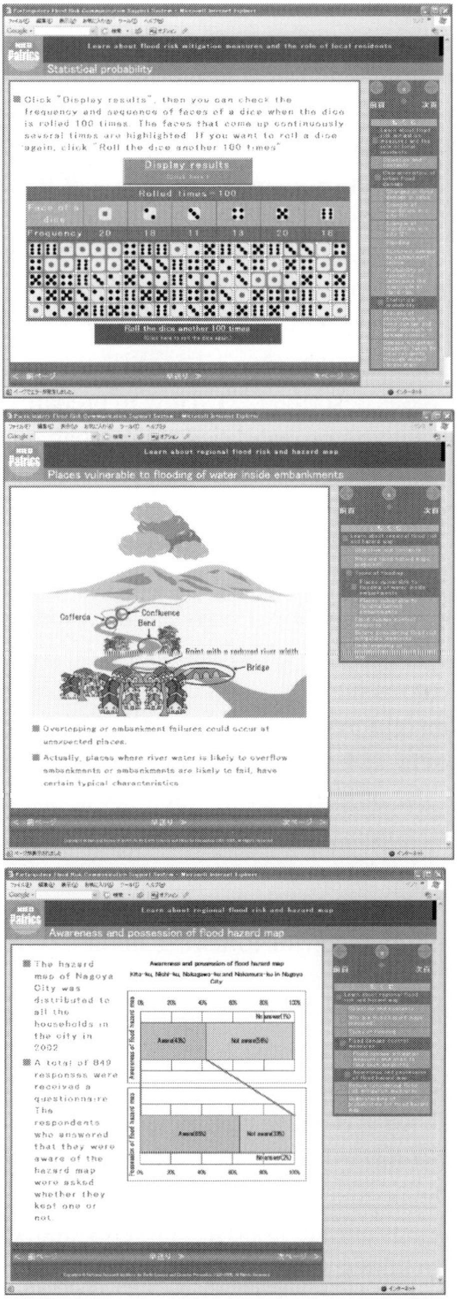

Fig. 7. Sample contents at Pafrics website (Originals are in Japanese).

4.2 Scenarios in relation integrated flood risk management

Presented here are the scenarios under "flood risk reduction measures and the role of local residents" and "Local flood risks and hazard maps".

(1) Sample scenario 1: Learning about flood risk reduction measures and the role of local residents

The objective of the workshop under the theme of this scenario is to "prepare for flood damage that could occur at any time". The workshop consists of 1) "Characteristics of urban flood damage", 2) "How to prepare for great flood damage" and 3) "Measures for reducing flood damage through the co-operation of local residents". Twenty-four contents constitute the workshop. Under 1) "Characteristics of urban flood damage", the structural vulnerability of cities to flood damage is described. 2) "How to prepare for great flood damage" educates participants on probabilities and probable precipitation, which are important when considering flood damage, and presents examples of flood control measures and flood damage prevention measures taken by flood affected residents based on the results of questionnaire surveys described in Part II. 3) "Measures for reducing flood damage through the cooperation of local residents" lists not only conventional physical measures but also flood damage control measures taken through the cooperation of local residents. The contents are designed to achieve the objectives of the workshop and provide participants with an insight into flood risk communication.

(2) Sample scenario 2: Local flood risks and hazard maps

This scenario has the objective of learning about hazard maps, one of the most important soft measures of flood risk reductions. The workshop includes 1) "Understanding the background to the preparation of the hazard map that has been distributed", 2) "Understanding the allocation of flood risk reduction roles", 3) "Requirements prior to taking flood risk reduction measures" and 4) "How to use hazard maps". The workshop has 18 contents. In 1) "Understanding the background to the preparation of the hazard map that has been distributed", types, mechanisms, and topographic conditions of flooding are explained. 2) "Understanding the allocation of flood risk reduction roles" shows that many of the flood damage control measures can be taken under normal conditions and that hazard maps are useful for damage prevention under normal conditions. The results of questionnaire surveys on the awareness and ownership of hazard maps among flood affected residents are presented. In 3) "Requirements prior to taking flood risk reduction measures", the magnitude, frequency, and probability of flood damage are studied, and explanations are given about uncertainty, preparation process, and differences in presentation of hazard maps and from inundated area maps. 4) "How to use hazard maps" describes considerations when using hazard maps and how to use them. These contents are designed to achieve the objectives of the workshop and enable participants to understand the importance of observing nature of flood risks in local communities under normal conditions.

5 Closing Remarks

The system is now being evaluated through verification tests at a number of disaster prevention workshops in Japan for local residents to find ways to make it easier to use. The knowledge and information required for effective risk communication have been gathered and compiled in this system through cooperation with local residents, disaster prevention NPOs, and other concerned groups. The system performance will be enhanced by strengthening the cost-benefit analysis function through the addition of GIS functions and incorporating the latest research results, and by improving the method of content presentation. The system will also be extended so that it can be used for risk communication regarding other types of diasters such as land-slides, sediment and seismic ones.

References

Satoh, T., T. Fukuzono, S. Ikeda, and K. Seo (2003): Development of Participatory Flood Risk Communication Support System, *Proc. of Annual Meeting of Japan Society for Natural Disaster Science*, 223–224 (in Japanese).

Takao, K., *et al.* (2003): Do perceived flood risk and fear of floods directly affect behavioral intention of preparedness against floods?: The case of the Tokai Flood Disaster in Japan, *The 3rd IIASA-DPRI Meeting (http://idrm03.dpri.kyoto-u.ac.jp/Paperpdf/47takao.pdf.*

Takeuchi, Y., K. Takao, S. Shimokawa, T. Satoh, T. Fukuzono, and S. Ikeda (2005): Education Effect of Learning Support Tool of Flood Risk Literacy, *Report of National Research Institute for Earth Science and Disaster Prevention*, **67**, 63–71 (in Japanese).

A better integrated management of disaster risks: Toward resilient society to emerging disaster risks in mega-cities, Eds., S. Ikeda, T. Fukuzono, and T. Sato, pp. 213–224.

Flood Risk Communication with Pafrics

Yukiko Takeuchi and Isamu Suzuki

1 Introduction

An approach of integrated flood risk management is needed to protect society against flood disasters. Residents, local communities, and government agencies will all be involved in the flood risk management. For effective flood risk management, risk communication between residents, local communities, and government agencies will be necessary. Therefore, the disaster prevention knowledge available to each of these groups should be improved. To support risk communication to prevent flood disasters, the Research on Social Systems Resilient to Natural Disaster project team at the National Research Institute for Earth Science and Disaster Prevention (NIED) has developed a new system: the Participatory Flood Risk Communication Support System (Pafrics). Pafrics includes three support functions; i.e. a learning support system for flood risk literacy, making a choice of flood disaster mitigation measures, and supporting to hold workshops. This chapter is concened with the examination of effective Pafrics from the questionnaire survey result to the participant by using Pafrics in the university lectures and the workshops in local communities. Examines are three; the educational effectiveness of the Pafrics learning support system for flood risk literacy, the workshop conducted using Pafrics, and the use of web version of Pafrics at the residents meeting.

2 Effectiveness of Pafrics for improving Flood Risk Literacy

2.1 Case of university lectures

The purpose of the learning support system for flood risk literacy is to help users better understand flood risks. To test the effectiveness of this learning support system, its application in a university lecture has been examined.

The lecture was held in 2004. There were 93 participants who ranged from second-year to fourth-year students of the university. The lecture title was "A lesson on integrated flood risk disaster prevention and hazard maps", and the goal was for the audience to fully understand flood mechanisms, the natures of disaster risk and as well as the use of hazard maps as one of the important soft measures to prevent disaster damages. For the 60-minute lecture, 21 slides were chosen from the learning support system for flood risk literacy in Pafrics.

The lecture first dealt with flood prevention measures and explained that the frequency or probability occurring floods of floods is an important factor to implement flood prevention measures. Then explained the purpose of hazard maps and the process through which they are created. As teaching materials, Pafrics and the Shonaigawa -Shinkawa flood hazard map which are prepared for Nagoya City were used. Copies of the map (reduced to A4 size) were distributed to the participants.

A 15-minute questionnaire survey was conducted before and after each lecture to check how well the lecture was understood. The questionnaire also asked about the respondents' background information and flood disaster experience.

2.2 Method

The questionnaire results before and after the lecture was designed to examine whether the respondents gained a better understanding of flood disasters and a basic understanding of the main points concerning the distributed flood hazard map.

The questionnaire focused on risk information provided by the map. To measure basic understanding of the flood risk information from the Shonaigawa-Shinkawa flood hazard map, the questionnaires asked about landform conditions that affect whether a flood disaster is likely to occur, a flood scale and its speed, the occurrence frequency of flood disasters, and the uncertainty involved in the simulation used to prepare the hazard map.

The questionnaire asked each respondent to choose the best response from five choices: strongly disagree, disagree, do not know, agree, and strongly agree.

2.3 Results and discussion

The differences of the participants' understanding about the pre- and post-lecture on flood risk are shown in Fig. 1. There was not a large change in the responses regarding the flooded area. However, the responses for "usage attention", "method of map reading", "map making aim", and "original memo about disaster prevention", and another item provided by Pafrics were significantly positive. Those for "usage attention" increased drastically. These changes are probably due to the explanation of how the hazard map was produced and the fact that it was distributed among the students and explained in detail through Pafrics.

The changes in flood risk information before and after lecture are shown in Fig. 2. The degree of understanding about "featuring landform possible flood disaster", "flood disasters frequency", and "the uncertainty involved in the simulation" increased. The students' understanding of "a flood scale and its

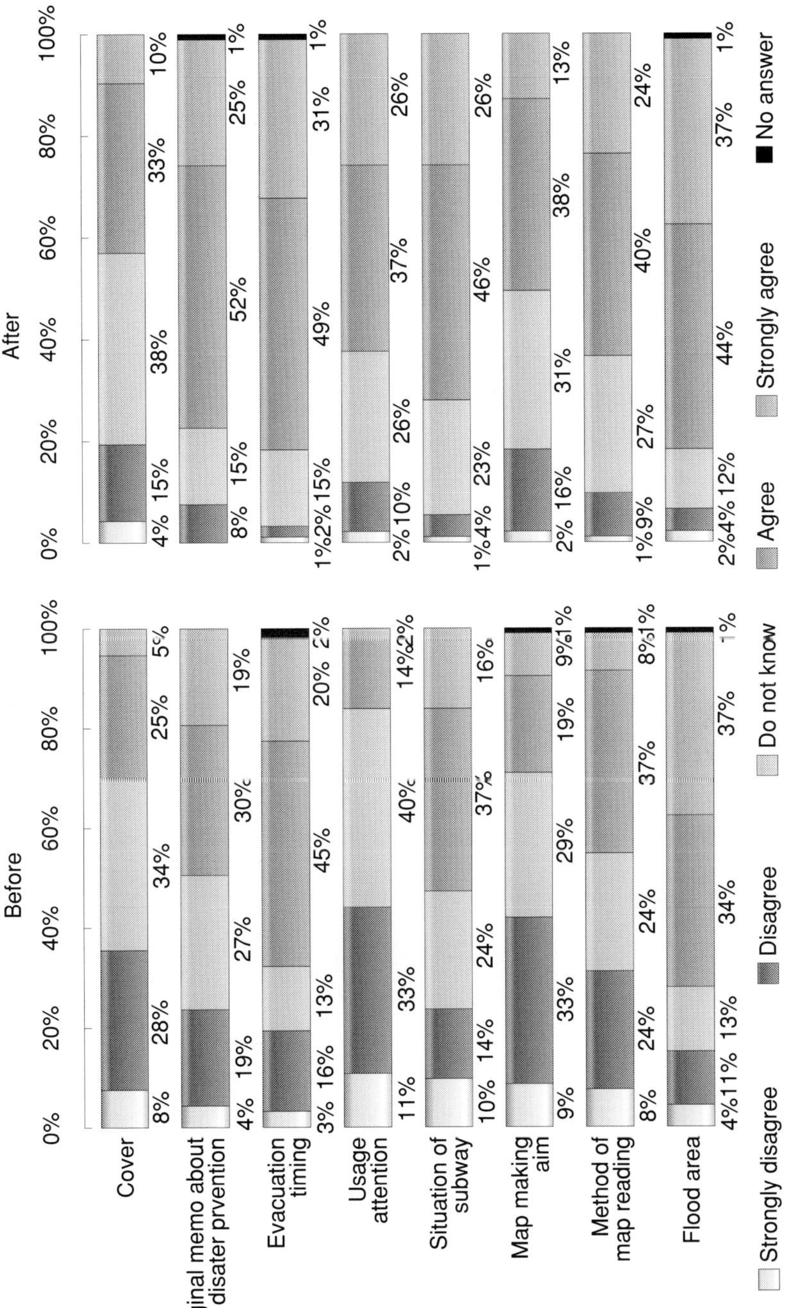

Fig. 1. Comparison of the knowledge of participants before and after the lecture on flood risk information.

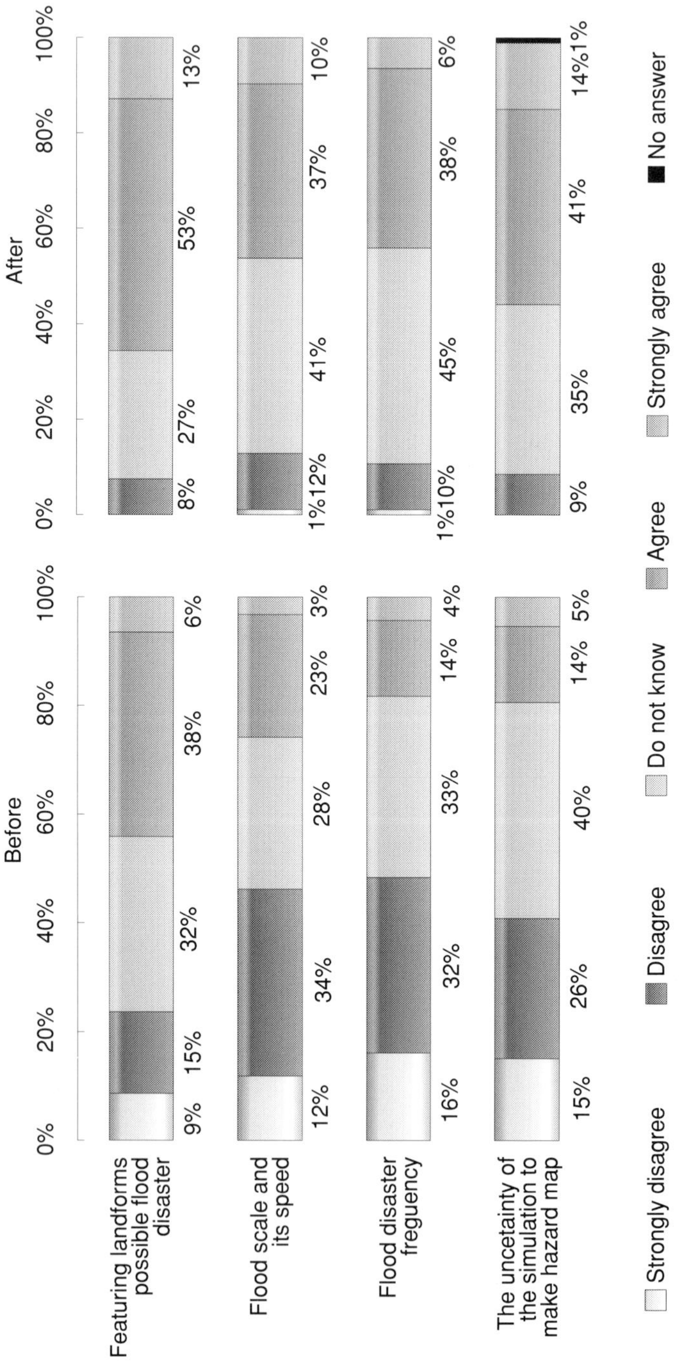

Fig. 2. Pre- and post-lecture flood characteristic understand by the participants.

speed" and "the uncertainty of the simulation to make a hazard map" deepened considerable. These changes also appear to have been partially the effect of the use of Pafrics to explain these risk issues.

The results from the questionnaires suggest that the Learning Support System for Flood Risk Literacy is useful for disaster prevention education. This system can thus be used to provide education in places where knowledge about namely flood risk is needed, such as in schools or within local communities. The results also suggest that much of the overall educational effect comes from Pafrics itself. The Learning Support System for Flood Risk Literacy provides a content collection which allows users not only to learn about a particular area but also to gain a general knowledge about flood risk. For risk communication to take place local areas, specific area content must be prepared. However, area information varies in its form, precision, and available quantity. During the system construction, it was thought that each local workshop organizer would be able to register the necessary local area content on Pafrics. It was decided, though, that the content has to be registered in a way that ensures reliability, objectivity, and public access, while also taking into account privacy and copyright concerns. Therefore, registering local area content creates many problems. We think that it is best to prepare local area content individually possibly by residents, themselves at present.

Integrated communication about risk is an effective way to implement risk management. In the United States, the National Research Council (1989) defined risk communication as "an interactive process of exchange of information and opinion among individuals, groups, and institutions". The smaller the differences in the knowledge held by all parties and the greater the awareness of differences, the smoother communication is likely to be. To enable smooth flood risk communication, Pafrics may contribute to even out the differences in knowledge of individual involved in flood risk management.

3 Pafrics Workshop

3.1 Case of river management

During the process to develop a river management plan, which included the participation of the local area residents, our project conducted a Pafrics-based workshop on flood risk communication. This workshop was called the Tokigawa-Shonaigawa Korekara Project (the Korekara Project).

Tokigawa-Shonaigawa is an urbanized river in the Chubu area of Japan. It flows into Ise Bay after crossing the Noubi plain from the Gifu Prefecture ravine area. In Gifu Prefecture, it is called the Toki River; while in Aichi Prefecture it is called the Shonai River. Urbanization in the basin has proceeded rapidly since the 1950s, and about 4,000,000 people now live there.

The purpose of the project was to create a river management plan for the Toki River and Shonai River based on discussions between experts, city representatives, and residents. The revision of the law concerning the river management was part of the project background. Local area residents (from both the upper and the lower reaches of the river), government agencies, non-profit organizations (NPOs), and experts in this field participated in the project.[1]

The Korekara Project consisted of the basin conference, a local area discussion meeting, and then a river administrator and Tokigawa-Shonaigawa meeting. The river management plan was developed in five steps:

1. Determination of the project direction
2. Ranking the priority of problems
3. Getting goals of the river management goal
4. Examination of the river management plan
5. Decision regarding the river management plan

The basin committee was established to hear opinions from river-management experts. The first committee meeting was on March 3, 2003. The river administrator and Tokigawa-Shonaigawa met to hear opinions from the Aichi and Gifu prefecture governments and communities within the basin. The first meeting was on February 4, 2004. An area residents meeting led to citizens' opinion-exchange meetings, open houses, and a "Kurumaza[2]" meeting. The open houses were display events where visitors were directly asked about their opinions regarding the information of the river; they were held at the town hall, in local shopping centers, etc. Opinions were gathered and put together as the Korekara Voice or the Report of the Korekara Project, and, consequently, were reflected in the river service plan. The "Kurumaza" meeting was to hear the concerns and specific needs of people living in the basin. This meeting was carried out in a face-to-face style. Ten citizens' opinion-exchange meetings were held to discuss and exchange local information and proposals regarding river maintenance. The participants included residents,

---

[1]The participants, apart from residents, were from the Ministry of Land Infrastructure and Transport (the river administrator), the Rescue Stock Yard (an NPO based in Nagoya City), a university faculty, and the Disaster Prevention Research Group, National Research Institute for Earth Science and Disaster Prevention (NIED), Japan.

[2]Kurumaza-meeting is open exchange meeting public.

representatives from resident groups, people who worked in the basin, and those who had some interest in the management of the river. They exchanged opinions regarding four themes: having a beautiful river that supports various ecosystems; utilization of virgin landscape, history, resources, and space; "To tell the large stock of experience about the river, make the river a place for relaxation"; and how basin residents and the local communities can think about flood control and prevention. Activities included collecting opinions and ideas regarding the river plan and the joint enterprise plan, making a map, and walking along the river. The consulting staff and NPO provided management and a facilitator for each citizen's opinion-exchange meeting.

3.2 Case of the workshops at local community

Our project team conducted two workshops supported by Pafrics for flood control and prevention work groups as part of the Korekara project. The participants were river basin residents and a river administrator. A member of the consulting staff, an NPO, or the NIED project team served as the facilitator. Each workshop lasted about 30 minutes and was based on one of two themes. After that, a participant takes part in group discussion referring to the workshop and gathered opinions, and forms an opinion.

3.2.1 First workshop: Flood control measures

The first workshop, on flood control measures, took place in Nagoya City on August 1, 2004. The theme was how residents can prepare against severe large-scale flood damage. The participants were six local area residents, two river administrators, three NPO members, and three NIED project members. Basis maps, vellum paper, memo pads, and Pafrics were used as learning materials.

Pafrics used workshop scenario 1, "learning about the flood damage risk reduction plan and the role of area residents", which was described in Chapter 13. After the learning session, the participants discussed the theme to identify problems that could occur during normal times; such as an emergency, a disaster, and the recovery phase and possible countermeasures to each problem.

3.2.2 Second workshop: Hazard map

The second workshop, on hazard maps, was also held in Nagoya City on October 3, 2004. The purpose of the workshop was to explain what information can be obtained from a hazard map, how a hazard map can be used, and to discuss local area community awareness of disasters and preventive action. The participants were six residents, four NPO members, and four NIED project members. The learning materials were a local area hazard map, vellum paper, memo pads, and Pafrics.

Pafrics used workshop scenario 3, "learning about flood risk and hazard mapping in the local area", which was described in Chapter 13. After the learning session, participants extract and categorize the information they could, and could not understand from the hazard map. They discussed ways to provide flood risk information and the measures a local person in charge of protection against disasters in the area should take to reduce the risk of area damage.

3.3 Results and discussion

Risk communication has been defined as "an interactive process of the exchange of information and opinion about flood risk among individuals, groups, and administratire". Conventional risk transmission has been more of a one-way information communication process from experts to local area residents. The current approach to risk communication, though, emphasizes that opinions and information should be exchanged in ways that encourage clear expression and mutual confidence. Therefore, three aspects of risk communication require careful consideration: 1) instead of information being provided in one direction only, there should be a two-way exchange of information and opinions; 2) an expert must be willing to provide enough information to non-experts for them to make reasonable choice among proposed measures; and 3) building a trusting relationship.

The workshop was evaluated with respect to these three aspects. There are four reasons why this was necessary. First, at previous meetings where river administrators and river-management experts stated an intended policy and provided information in a one-sided manner to local residents, they were unable to persuade the residents to support the policy. When information was shared in a two-way exchange and the resident's or opinions were expressed, both sides gained a deeper understanding of the present situation and the necessity of the river management plan. Second, awareness of flood risks was improved through mutual dialog. Specific risks recognized by some participants were sometimes apt to be neglected by others. Also, the river administrator and the experts could explain risks which residents were unaware of. Third, an effort was needed to build up a trusting relationship through the mutual understanding of participants. Although criticism, personal wishes, and problems are often expressed in open dialog, opposition could be managed and hopefully a consensus could be reached. Participants might established amity through such discussion. Fourth, an effort was needed to ensure the river maintenance plan reflecting the outcomes of discussions. Such outcomes were gathered together as a proposal of the residents and presented to the river administrator.

As for the risk communication, the facilitator played an important role.

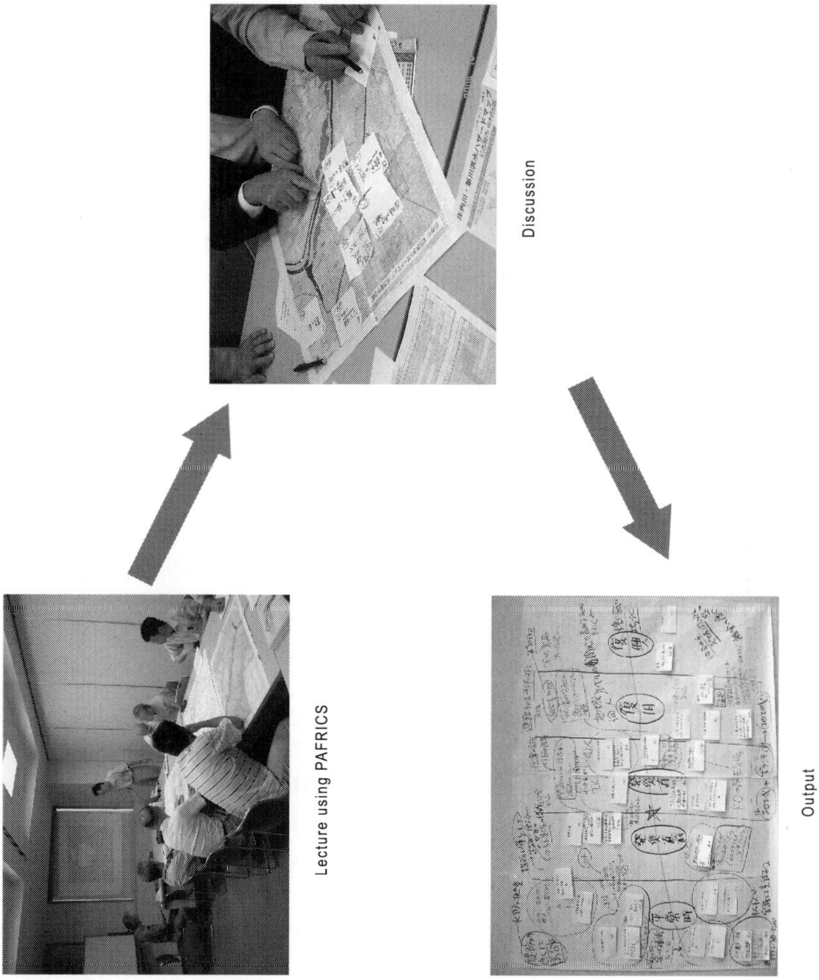

Fig. 3. Flow of Pafrics Workshop.

In addition to showing knowledge and experience, the facilitator took note of the main points of the participants' opinions and was able to use this middle ground to encourage consensus among the participants. In this way, the risk communication could proceed smoothly.

Out of the purposes of Pafrics is to promote this sort of smooth risk communication at workshops and assist the facilitator. The staff involved in this workshops says that compared to the former discussion, the workshops using Pafrics were more specific in terms of clarifying the issues of of river flood management. Throughout each workshop, residents showed a conviction that they should work on disaster prevention in the local community and not be overly reliant on NPO activities. There was also a desire to disseminate the knowledge and interests of residents to others. Although these results will be analyzed further to confirm the effectiveness of Pafrics, they acknowledged that Pafrics can promote risk communication and support workshop facilitators.

In the future, it is important that the residents become a mainstream to continue a discussion about river management plans and prevention about flood disaster.

The participants showed a deep interest in reducing flood risk and had considerable experience and knowledge during the workshops. However, the questionnaire results showed that they were resistant to becoming a mainstream leader in a locality. The effective use of Pafrics or other study tools, though, should enable residents to continue useful discussions on river management plans and flood disaster prevention measures in cooperation with NPOs and similar organizations.

4 Evaluation of the Pafrics Workshop Support System

4.1 The Pafrics workshop support system on the Web

When planning a workshop, in addition to deciding on a theme and finding an objective person to facilitate, it is necessary to prepare the workshop content. In addition, while the workshop facilitator plays an important role in making sure a workshop proceeds properly, a good facilitator can be hard to be obtained because of the preparation and knowledge needed. Therefore, even if people recognize a workshop is needed, it might be hard to realize. To alleviate this problem, the Web version of Pafrics provides various items needed to hold a workshop. For example, it provides a manual with information about the preparation content, the schedule, and the workshop theme. Based on a report on workshops held using the Web version of Pafrics, the workshop support system of the Web version is discussed below.

4.2 Evaluation of workshop support system

This report was provided by NPO which conducted a workshop using the Web version of Pafrics. This workshop used scenario 1 (described in Chapter 13) of the Web version of Pafrics. The workshop topic was the fear of flood disaster and awareness of the role of residents in prevention. There were 25 participants. After the workshop, the NPO members completed a questionnaire about the Web version of Pafrics. The questionnaire (free description) was chosen from the web version Pafrics.

The questionnaire results revealed that in one part of the workshop it was difficult to understand the content and in one instance there was a problem with the color and size of the screen. However, the results pointed out the possibility of holding a workshop without preparatory practise. Moreover, it is feasible that the workshop can be held even if the facilitator has no experience. From these answers it was suggested that the existence of a manual has extensively supported the facilitator to organize the workshop. Besides, it is though that, implementation of workshop without practising show that there is a certain progress of the workshop support system of Pafrics. It suggests that the improvement of the manual leads to the perfection of the workshop support system.

5 Concluding remarks

Based on the questionnaire survey, we evaluated the educational effectiveness of learning support system for flood risk literacy, the effectiveness of workshops support, and the usefulness of Pafrics as a workshop support system. The results from the lecture suggest that the learning support system for flood risk literacy is useful for the understanding of risk information and build up the workshop for disaster prevention education. In addition, Pafrics appeared to promote two-way risk communication and aid the facilitator during the workshops. Finally, the questionnaire responses from the NPO which held the workshop using the Web version of Pafrics indicate that the support system of Pafrics was useful. The performance of Pafrics will be improved with repeated use. To enable support of many other types of risk communication in the future, Pafrics will be extended to reflect more opinions and operations in terms of other disasters.

References

National Research Council (1989): Improving Risk Communication, National Academy Press.

Takeuchi, Y., Takao, K., Shimokawa, S., Satoh, T., Fukuzono, T. and Ikeda, S. (2005): Education Effect of Learning Support Tool of Flood Risk Literacy, National Research Institute for

Earth Science and Disaster Prevention, No. 67, 63–71 (in Japanese).
Tokigawa - Shonaigawa Korekara Project official web page (2005): http://www.cbr.mlit.go.jp/shonai/body/4.html

Index